CASE STUDIES IN

CULTURAL ANTHROPOLOGY

GENERAL EDITORS

George and Louise Spindler

STANFORD UNIVERSITY

ROYAL BLUE

The Culture of
Construction Workers

ROYAL BLUE

The Culture of Construction Workers

By

HERBERT A. APPLEBAUM

HOLT, RINEHART AND WINSTON
NEW YORK CHICAGO SAN FRANCISCO DALLAS
MONTREAL TORONTO LONDON SYDNEY

To Earl Kostuk, my friend and coworker
And to all the other construction workers
Who daily build our nation

Also dedicated to my children:
Stephen, Michael, Robert, Eric,
Amanda, Aleta, Alexia

Library of Congress Cataloging in Publication Data

Applebaum, Herbert A
Royal blue, the culture of construction workers.

(Case studies in cultural anthropology)
Includes bibliographies.
1. Construction workers—United States. 2. Construction industry—United States. I. Title. II. Series.
HD8039.B92U54 306'.3 80-18487
ISBN 0-03-057309-2

Printed in the United States of America
1 2 3 4 144 9 8 7 6 5 4 3 2 1

Foreword

ABOUT THE SERIES

These case studies in cultural anthropology are designed to bring to students, in beginning and intermediate courses in the social sciences, insights into the richness and complexity of human life as it is lived in different ways and in different places. They are written by men and women who have lived in the societies they write about and who are professionally trained as observers and interpreters of human behavior. The authors are also teachers, and in writing their books they have kept the students who will read them foremost in their minds. It is our belief that when an understanding of ways of life very different from one's own is gained, abstractions and generalizations about social structure, cultural values, subsistence techniques, and the other universal categories of human social behavior become meaningful.

ABOUT THE BOOK

With this case study on construction work and workers in the United States, we now have three ethnographic studies in the Case Studies In Cultural Anthropology series on different segments of North American industry. They discuss longshoremen in Seattle, railroading, and now, construction. When we began the series in 1960 with case studies on Palau, the Tiwi, Tepoztlan, the Bunyoro, and the Cheyennes, we declared our intention to include in the series ethnographies from any part of the world where anthropologists worked, including our own society. We regard the principles, concepts, and methods of cultural anthropology as universally applicable. There are now 20 case studies published on segments of North American culture. They are listed at the end of this case study. We hope to publish other studies reaching into as yet ethnographically untouched areas of American life.

In *Royal Blue,* Dr. Applebaum gives us something unusual—not merely because it is an ethnography of construction work and workers, but because the study combines a literary-humanistic, even poetic view in combination with a great deal of substantial, solidly analyzed observation and factual information. The phenomena studied are analyzed from both the cultural and structural-functional viewpoints. Sociological and anthropological concepts are combined in a smooth analytic meld.

The duality of humanistic-literary, anthropological-sociological views and approaches gives a nonsectarian character to this case study which should make it attractive to a wide audience. And, although *Royal Blue* is an ethnography of construction work and workers in four construction sites where Dr. Applebaum was an observing participant, it is contextualized in the broad frame of reference of the construction industry in North America. One of the salient features of an emerging industrial anthropology will be attention to regional and national (and indeed, international) systems of which the localized phenomena studied are a part.

ABOUT THE AUTHOR

Herbert Applebaum has been in the construction field since 1956, serving on construction projects as a superintendent, project manager, and project engineer. He has also been employed in management positions—director of construction, chief estimator, contract administrator, and vice-president and president of his own construction firm. He was also a consultant for three bonding companies in the construction industry. He has taught classes in construction management in New York City. He has also taught classes in anthropology, sociology, and economics on a part-time basis at Mercyhurst College and the Behrend Campus of Penn State University in Erie, Pennsylvania; at Monroe Community College in Rochester, New York; and at Erie Community College in Buffalo, New York. He has been teaching for 13 years, mostly in the evening and in adult-education programs. He is presently employed as a project manager for a construction firm in western New York State. The data for this study was accumulated on four large projects in this region over a four-year period, from 1975 to 1979. The company he worked for was ranked 143 among the 400 largest general contractors in the United States in 1974.

Dr. Applebaum says of himself:

> While I was project manager on the Amherst sewage Treatment Plant project, some very dramatic events took place at work and I tried to write some literary pieces about them. One result of this effort was "The Thousand-Yard Pour," included in this book as Chapter 1. I also started to write poems about construction work and workers. I showed them to my coworkers and was encouraged by their response. It occurred to me that I was in the unique position of being an anthropologist employed where the action was—that if I studied construction workers I would need no artificial entry into some "community" or organization. I was already a member of the group.

Herbert Applebaum received his Ph.D. in Anthropology at the State University of New York at Buffalo in 1979.

George and Louise Spindler
General Editors

Phlox, Wisconsin

Preface

As anthropologists increasingly study complex and modern society, they will inevitably investigate work and occupations which link individuals to social institutions and provide social identities. Work and occupations are vital units of study regardless of what model one may employ. The world of work is important if we are looking at the process of socialization; if we wish to study class and status; if there is concern with social role; or if one takes an institutional approach to culture.

This book deals with the construction industry. Construction is the largest single industry in the United States, employing more than 4 million workers and accounting for 13 percent of our Gross National Product. It is not larger than all of manufacturing, but it is larger than any single branch—larger than auto, electrical equipment, rubber, steel, chemicals, oil, textiles, and clothing. It is an industry responsible for 50 percent of our capital growth. Thus, it is a crucial part of our society.

The main emphasis of this book is on the technological and social organization of the construction industry which fosters a particular pattern of work behavior. The behavior and the values that accompany it stress independence and autonomy of the construction craftsman. He owns his own tools and controls the work process. Construction workers enjoy a high degree of job satisfaction based on their autonomy, seeing the physical evidence of their labors and performing their work in gangs and teams where they can enjoy personalized, collaborative efforts with fellow workers.

The social organization of the construction industry is based on a number of unique features. One is the localized nature of the industry. Every structure is fixed to a local site and thus, all construction work which is performed on the site is, of necessity, decentralized. Secondly, each structure, especially the heavy commercial and industrial construction, is unique. This precludes the utilization of mass production, which is associated with repetitive work tasks and bureaucratic administration. These two factors, along with the long-term nature of the productive process and the unpredictability of weather, reduces the degree of control that can be imposed on the construction process. Thus, uncertainty and imprecision are norms for the industry. The result of all these elements is an atmosphere on the construction job site of informality, face-to-face relationships, and loose, personalized supervision. It is an environment which generates a human quality to the work that is reinforced by the hand technology and the reliance on craft skills to create the product.

The theoretical orientation of this case study is based on the idea that the social organization of construction work and its consequent behavioral patterns among construction workers lead to a distinctive occupational subculture. It is a type of culture similar to that found among other roughneck occupations— longshoremen, seamen, loggers, miners, pipeliners, truckers, and other "independents" among the blue-collar trades. It is also similar to craft cultures such as printers or glassblowers.

The dominant culture in the United States is strongly influenced by the bureaucratic mode of organization, as exhibited by large corporations, factories, and huge office complexes. Construction workers are not part of this type of work environment. Yet they are not divorced from the culture of the larger society. They are influenced by the mass media. They live in urban areas. They are, in part, a product of their ethnicity, religion, race, sex, education, and geographical region. They are part of an integrated economic structure, in which they share many things with other Americans as consumers, home owners, and taxpayers. Nevertheless, they are a discrete entity within the framework of the larger society. They are bound to one another in work and family networks that do not include others, and they project relationships, values, and norms from their work into their nonwork lives.

Work and occupations constitute focal points in the lives of people in modern society. Anthropologists are now turning their attention to modern cultures and studying occupational and other groups. Hopefully, these studies will produce a large number of industrial ethnographies, from which there may emerge as a subfield within the larger discipline of anthropology an "anthropology of work."

I want to express my thanks to George and Louise Spindler and to David P. Boynton, without whose help and consideration this book would not have been published. I also wish to acknowledge the help I received from my colleagues at the State University of New York at Buffalo—Erwin Johnson, Terrence Tatje, and Keith Otterbein, who read the manuscript and offered sound advice and critiques. I would like to thank Roy Kaplan for his help, and Jack Rollwagen, who made the right suggestion at the right time and place to help this book see the light of day. I am especially grateful to my wife, Mika, who maintains her daily belief in my work. I also want to acknowledge the contribution of my parents, Rose and Herman Applebaum. I want to acknowledge the help and advice of Robin Gross of Holt, Rinehart and Winston.

Contents

1/The thousand-yard pour

The day of the thousand-yard pour arrived.

On the morning of the pour, everyone was up and out on the job early. It was going to be a long day and Earl wanted to get started at the crack of dawn. In fact, we were on the job before dawn. It was still dark. The immense and timeless sky bent over us in its blackness and we couldn't tell what kind of a day it would be. Everyone's face was lifted skyward, sharing a collective hope for a good day, meaning one without rain. The resident engineer was there, frowning importantly.

It was that time of the morning when a mysterious silence fills the air, as if the world hovers in restless anticipation. The men talked in quiet tones and smoked cigarettes that glowed like fireflies. Despite the uncertainty, they waited patiently. Some talked about the pour; others joked; some exchanged news about yesterday's baseball games; a few discussed food and offered each other favorite recipes; and still others talked about politics.

As the first sign of light cracked the horizon, cobblestone clouds, the color of gray slate, appeared in the sky. They were foreboding clouds that augured a stormy day. They did little to raise our spirits, but we had no time to worry about it because we could hear the first truck muttering down Tonawanda Creek Road, about a quarter of a mile away.

Earl yelled, "Here it comes!"

The pour was under way.

Before continuing with the story I have started, I want to explain about the construction workers who poured a thousand yards of concrete in a single day, a prodigious feat, involving many weeks of planning and the solution of a number of difficult problems.

It was the summer of 1976. In the summer, construction reaches its peak. Rain and bad weather are less frequent and work can proceed without interruption. Each day, structures become taller and alter in shape, and new accoutrements adorn rising buildings. The project site changes—hills disappear, new embankments arise, brush and weedy growths are swept away, streams and waterways are bridged. Everywhere dust is blown and swirled by noisy, lumbering construction machines and trucks that crisscross the site in seemingly chaotic patterns. Change takes place unexpectedly, wondrously, as if by magic. But there is no magic—only the palpable efforts of men working and planning together in teams and gangs.

We were working on the new sewage treatment plant for the town of Amherst, New York, that summer. We had 13 different structures to build: a pump station; a chlorine holding tank; several clarifiers in which sewage is oxygenated; a grit chamber, in which solid particles are removed and deposited in dumpsters; an equalization tank, where sludge sinks to the bottom and is then scraped into troughs and pumped to an incinerator building where it is burned; and a variety of collection tanks.

The equalization tank had a floor slab that was four feet thick and contained 2000 cubic yards of concrete. It was as long as a football field and about half as wide. The men on the job decided they could pour that floor in two days, each day placing 1000 yards of concrete on half the area. I was the project manager, and when the men came to me to tell me what they had in mind, the idea seemed awesome. I had never taken part in such a feat and had little idea of the details and problems it involved. I discovered that weeks of careful planning were neces-

Figure 1. Author (right) with a grade foreman named Frenchie. Behind us is a backhoe, a Dynahoe 200. We used this machine to dig for sewer and waterlines. Picture was taken on the Elm–Oak arterial, a road and bridge job in downtown Buffalo. Cost: $12,000,000. (Photo courtesy of The Buffalo Evening News.*)*

Figure 2. Earl Kostuk, construction superintendent and co-worker, to whom this book is dedicated. (Photo courtesy of John Kaiser, Roy Crogan & Son.)

sary to account for innumerable maddening details. We prayed that weather would be on our side the day of the pour in spite of "Murphy's Law"—whatever can go wrong, will.

Earl, the project superintendent, was the first to tell me that he and the men decided on the thousand-yard pour. Earl was the best superintendent the company

Figure 3. The result of the "Thousand-Yard Pour." The entire floor, which is four feet thick, was poured in one day. This is the floor of the equalization tank at the Amherst sewage treatment plant, Amherst, New York. (Photo courtesy of John Kaiser, Roy Crogan & Son.)

had. He was an honest, straightforward, no-nonsense man. He had an unswerving gaze and looked directly at the person he addressed. He was a carpenter by trade, but over the years he had learned a great deal about every trade and could give direct leadership to most construction operations. In his working style, he always sought the consent of the men he asked to do something. He would discuss the various ways of performing a job and get the craftsmen to agree on the best method. He had an uncanny memory. If someone told him something and then tried to change or deny it, he would catch that man in a flash, and repeat to his face, word for word, exactly what he said, when he said it, and under what conditions. Earl wasn't remote from the men he led. He drank and kidded with them, and went hunting and to sporting events with them. He spent most of his social life with other construction workers who had become his friends over the years.

One day, in the shanty, I listened as Earl was discussing preparations for the pour with Iggy, his labor foreman, and Bob, one of our surveyors. They were considering the use of a concrete pump or a crane to place the concrete.

"The pump works fast," Iggy said, "but it's always breaking down. Don't matter if it's old or new. I've never seen one go through a whole pour without something happening."

One of the ironworkers, who was setting reinforcing steel in the forms that would hold the concrete, came in to see Earl to go over a steel drawing and get a clarification of what was shown. Earl had to interrupt the conversation with Iggy while he explained to the ironworker what the drawing meant. Returning to the subject, he told Iggy, "I know pumps break down. But we've only got ten hours to make that pour. And the pump can put out more than a crane. We gotta make sure the pump is checked out and serviced in the yard before it comes out to the job."

Iggy was a realist. He knew that no concrete pour ever comes off without a hitch, and he was trying to emphasize that point with Earl. Iggy was like Earl in character and personality, which is why he was always the labor foreman on Earl's jobs. He had a sense of his own dignity and pride that everyone recognized.

Iggy was also a fatalist. His two favorite words whenever he discussed philosophy or religion were *chance* and *fate.* He believed that through chance and fate our lives were determined and woven together, and a force or God behind it all was hidden from our vision and knowledge. So he did his work each day based on the experience and knowledge he had gained over the 20 years he had been in construction. But whenever some accident took place or a man was hurt through some mishap, he would point it out and remark that no matter how much man knew or accomplished, he was only a small atom in the gigantic "web" of things. Iggy was also fond of the word *web.*

Earl continued the discussion about the pump and the crane:

"Why can't we use both? We can have a crane on one side and a concrete pump on the other. This way, if the pump breaks down, we'll have the crane to finish the pour."

Iggy agreed, "We can have two separate crews. Let 'em fight it out which gets done first on their half. Maybe we can take bets on it."

Construction men will gamble on anything. They love to bet. They enjoy card playing during lunch hour. They are big customers for Las Vegas junkets. Many of them come to work each day with fistfuls of horse racing tickets. One of the labor foremen owned a racehorse. Its name was "Green Gertrude," after a small bulldozer which the men nicknamed "Gertrude" and was painted green. The men showed their loyalty to the foreman by betting on his horse whenever it ran. The horse had one of the worst records, but the men supported their work mate and kept betting it despite their losses. They kidded Johnny, the horse's owner, through all its difficulties, but they never failed to bet on it. Finally, the horse won a race. The joy on the project that day was unparalleled.

A week after the decision to go ahead with the thousand-yard pour, the field representative for the concrete supplier was invited to our field office. Earl spoke to him.

"Look, Walter, we're going to pour the equalization tank in two days, a thousand yards each day. I figure it will take about ten hours a pour. Your trucks hold 12 yards. That means we're going to need 84 trucks for the day, or eight trucks an hour. That means we'll be using one truck every seven minutes. The question is: can you give us the trucks and concrete? No bullshit now."

"It'll tie up our entire plant for the day."

"So what? Look at the business you're getting. Call your boss, Walter. I've got to have a definite answer. Otherwise I'll have to find another supplier."

"Okay, okay, Earl. There's no problem. We can do it. I'll just need enough advance notice so I can tell our other customers they get no concrete that day. But you gotta tell me what day. And what about our trucks? Will they be able to get in and out without getting stuck?"

"Yeah, don't worry about that. I've got a turnaround area. The trucks will be able to dump and get right out. I'll be building a new haul road soon. As for the day, figure about three weeks. But I'll give you the exact day in a week."

The next day, Earl and I picked out two separate routes to either side of the equalization tank and laid out two turnaround areas for the trucks. When Iggy came into the shanty at lunch time, Earl told him, "Herbie and me picked out two roads and two staging areas for the pour. After lunch I'll show 'em to you. Herbie agreed to spend the money for two good haul roads so we don't get screwed in case it rains. I want slag and brick brought in. Track it in good with the dozer. Then I want both roads rolled so the water will run off. Get on it tomorrow."

About a week later, Pete, the concrete finisher foreman, was called to the shanty. Pete is a dark-skinned Sicilian-American who turns black in the summer. He has too many teeth for his mouth and they are always visible. His bared teeth, black eyes, and powerful body give him a menacing appearance. Even his laugh doesn't diminish his intimidating face. Iggy once told a story of how Pete showed up late for a card game one night. He walked in wearing a black cape, white gloves, and a black hat, and scared everyone in the room out of his wits.

Earl told Pete, "Look, we got a big pour coming up. In a few weeks. I'll need two crews—finishers, spreaders, laborers. I figure 14 in each crew. I'll give you the exact day in about a week."

"I'll call the hall, today."

"No screw-ups. Tell 'em what we need. We'll confirm definitely by the end of next week."

"Whadda ya want me to say, Earl? When I tell my men to come, they come. J-e-e-zus. It's no problem, I tell ya."

"Sure, Pete, but it's a big pour and the men better be there."

"We poured bigger, Earl. Don't worry, we'll do a job for ya."

Pete had his following of men who worked together regularly. Although he was working for a contractor, he functioned almost like an independent subcontractor. He lined up the men, instructed them beforehand about what kind of pour it would be, what tools they would need, how long the job would last, what the job conditions were, and when he expected them to be on the job—without fail!

The next week, Earl called all the jobs in town being done by his employer. He spoke to the superintendent on each project, told them he had a big pour coming up, and gave them a list of numbers and types of men he needed: operating engineers and oilers, carpenters and ironworkers, flagmen, laborers, and surveyors. By that time, he had selected the date of the pour. He asked for, and received, a definite commitment of men on the day he had picked.

Within another week, the carpenters had completed their form work and the ironworkers had set and tied the reinforcing steel that would be incorporated within the slab. Meanwhile, Earl and Iggy continued to check out details and make their plans.

They double-checked the concrete plant. They studied the condition of the haul roads. The equipment yard was called to check out the condition of the concrete pump. Pete was questioned about his men. The surveyors were asked to check the height, width, and length dimensions of the forms. All other trades, particularly the plumbers and electricians, were asked to make sure they had all their pipes and sleeves in place that were supposed to go into the slab. The resident engineer was requested to make a final check of all items connected with the pour, to be certain nothing was omitted. As the time for the big pour approached, the men began talking about it with more than a touch of anticipation and excitement.

And now, the special day arrived and the big pour was under way.

Some men put out their cigarettes. Others put on their gloves, gathered their tools, and adjusted their hats and boots. The men wore a wide variety of clothes. There were combinations of tee shirts and blue jeans; overalls over long-sleeved work shirts; overalls over bare arms; rust-colored jackets; overalls with stripes; white overalls; high boots, low boots, high leather shoes, tan shoes, black shoes, worn shoes; tattered clothes, new clothes; soiled hats, hard hats, no hats. There was no common uniform.

The first concrete truck came bumping on to the job site. A flagman at the road directed the driver and the truck rattled and whined its way toward the equalization tank. At the pour site, the truck was hollered and maneuvered into position under the crane. The driver creaked back the emergency brake and leaped down from the cab. He must have felt the pressure of the large throng of men waiting to go into action to perform their work. Since he was the first truck driver, all eyes were on him. The man went to the rear of his truck and threw a lever that started the giant drum that contained the sand, cement, and stone which, when mixed with water, becomes concrete.

When the concrete was ready, Tse Tse, an expert "bucket man" (who was technically a laborer) swung the chute at the back of the concrete truck over a large round concrete bucket into which the concrete was to be deposited. It would then be lifted up with a crane and swung to the men waiting to empty it and shovel it into place.

As was his practice, that morning Tse Tse was wearing his hat sideways. I've never found out how he was nicknamed Tse Tse or what it means. Nicknaming is very prevalent among construction men. Many of the names are terms of insult, but when used among work mates or friends they are considered terms of familiarity. Let an outsider use the same word to call a man and he may find himself in a fist fight.

Figure 4. Laborer vibrating concrete to make sure it fills every part of the forms. If this is not done well, when the forms are stripped, it will result in "honeycombing," a series of holes left on the surface of the concrete. (Photo courtesy of Constructioneer, *a trade magazine.)*

Tse Tse loved to kid around. It seemed as if every day he thought up something new to amuse the men. One time he brought in a mannequin and positioned it in a field on a small knoll and crossed its legs. From far away it looked like a nude woman. Some of the men were curious and approached to see what was going on, and as soon as they realized what it was they howled with laughter. Whenever Tse Tse cleaned out a concrete bucket with a water hose, anyone who came within range received a good dousing. He had white hair and I've seldom seen him shaven. His grisly face was usually deadpan, but his eyes sparkled as if he were laughing to himself over his next surprise for the men. His reputation as a tireless and efficient bucket man made him wanted by the foreman whenever any sizeable concrete pour was being planned.

As bucket man, Tse Tse had responsibility for giving signals to the concrete truck driver and to the crane man who would swing the bucket to the men. When he wanted the truck driver to fill the bucket, he showed him a fist with his thumb down. Then a lever was thrown and the concrete whooshed into the bucket. As the concrete neared the top, Tse Tse signaled the driver by holding his fist with his thumb straight out, which meant shut off the flow. Then, he attached the crane hook to the top of the bucket and signaled the crane operator to lift by showing him a fist with his thumb up.

The crane then swung the bucket over to a gang of men, who seized it from all sides. They pushed and pulled it to where they wanted it. A laborer yanked a lever at the bottom of the bucket and the concrete cascaded down making a slapping, plopping sound as it hit.

On the other side of the equalization tank, a truck was backed up to the concrete pump, discharging its contents into a large, square tub. Inside the tub, two large screws pushed the concrete into the discharge cylinders where it was slammed through into the discharge hose by means of two piston heads operated by a pump. Two lines of hose stretched hundreds of feet away, with two men at each end directing the flow of concrete. When the concrete arrived, a slump test was taken.

Let me describe a slump test. A tapered steel cone, open at both ends, is filled to the top. The cone is lifted and the concrete slumps from its own weight. The difference in height between the cone and the slumped concrete is recorded. Concrete is too stiff if its slump is less than two inches and too loose if the slump exceeds six inches. Inches of slump and strength of concrete are related. Thus, every batch of concrete gets a slump test.

The full light of day was now upon the men without their being aware of it. The smoke-gray light had stolen almost imperceptibly through the darkness. The air shone gray-blue and was faintly luminous with the day. The earth rose solemn and lonely-looking in the long field that stretched away from the equalization tank.

Things were going smoothly and, as early morning passed into mid-morning, the men became more vocal. They were soon exchanging insults and jokes and feeling comfortable with each other. They had to walk on the raised platform of reinforcing bars, but they moved with graceful assurance, their arms and backs displaying power, coordination, and skill. They were joyful. They were masters of their tools and materials. Of course, they couldn't control the weather. But that would not stop them this day.

About 11:00 A.M., a commotion took place at the concrete pump. I saw Earl and Iggy running in that direction and followed them. When I got there, I heard Art, a mechanic, say, "The pump is turning over, but it's not pushing any concrete out the other end."

Earl shouted, "Get Bobby. Fast!"

The situation was serious. The concrete was hardening in the hose lines. If the pump did not get fixed in 30 minutes, the concrete would solidify and that would be the end of the hoses and one half of the pour. Earl and Iggy had to decide how long to wait before uncoupling hundreds of feet of hose, which had been bolted together in seven-foot sections. It was a lot of connections to undo.

Many pairs of hands and eyes were busy at the concrete pump. Earl ordered the piston head off and the gear mechanism checked. Bobby (the master mechanic) had half his body twisted down under the motor, calling for tools, like a surgeon at an operation. It was taking too long. They couldn't wait any longer. Earl hollered to Iggy, "Pass the word. Break the connections on the hoses. Get the concrete out. I want everyone on it. Tell Monkey to round up as many men as he can at the pump station and get them over here to help."

The work area exploded with men pushing, uncoupling, shouting, cursing, and puffing. The couplings on the connections were hammered free. The pipe was lifted on the shoulders of one man and the concrete pushed out by another. Some, like Earl, seized a section of pipe alone, and with enormous strength, lifted it on their shoulders and shook out the concrete by themselves. Everyone was splattered with concrete, their faces grayed from the cement and their hands cold and bruised. The relaxed chatter of an hour ago, the proud, self-confident ring of voices, gave way to hard anger and frustration at the necessity of doing work over.

Bobby found the trouble. A new part was needed. He jumped into his pickup truck and headed for the supply house to get it.

Earl told Iggy and Pete, "Get all the hoses recoupled and connected to the pump. By the time we get that done, Bobby should be back."

When Bobby returned, he went right to work, fixing the pump. As I watched him, I was struck by the power and massive shapeliness of his hands. His knuckles were huge, and he had thick, sinewy fingers. His hands looked twice the size of an ordinary man's. He worked on the pump parts with incredible speed and sureness, even though he knew dozens of men were waiting and watching him. When he was done, he nodded to Earl, who gave the word to restart the pump.

We were still filled with uneasiness. If the pump started, would it push out the concrete? If it pushed the concrete, would it crap out again with the concrete in the hoses? Would the men have to go through the whole frustrating breakdown of the operation again?

The pump started and concrete filled the hoses. We could see the sections of pipe jump and stretch full as the concrete entered each section. Suddenly, the pump failed. Bobby went to work on the motor again while the rest of us checked our watches to see how long he was taking. We were thinking of the concrete hardening in the hoses once more.

I heard the pump going again. Then Earl shouted, "Hold it!"

Off to one side, I could see a connection that loosened. A wet stream of cement was crazily looping gray pellets which were hitting and sticking to the men. Iggy and two others were on the faulty joint in an instant, hammering home the lugs that held the pipe in place. As soon as they were done, they signaled Earl, who gave the word to restart the pump once again.

With that, a lever was thrown and the whirring of the motor crescendoed into the blue-gray air that seemed thick with the anger of the men.

When the pump started, one of the laborers rushed to the end of the pipeline and scooped out with his bare hands the semihardened concrete. This permitted the new, fresh concrete to come pouring through. But another loose joint popped its connection. It jumped concrete into the air and the pump had to be stopped. The joint was retightened and the pump restarted. This procedure took place over and over. Each stopping and starting was related to a joint that loosened closer and closer to the end of the hose line.

Finally, after an hour of adjusting, hammering, hollering, and cursing, all joints were solid. The concrete was now flowing out at the end of the two lines, which looked like two overfilled mouths gulping up gray, mealy concrete cereal which fed the forms fabricated by the carpenters.

The pour progressed steadily through the lunch hour, through the early afternoon, and then, about four o'clock, it began to rain—a hard, pounding, saturating rain. With a downward rush, a blanket of rain picked a million points in the freshly troweled concrete.

I was in the shanty with Earl and Bobby as the rain slapped at the shanty window glass. Earl slammed his palm on the plan table and bellowed, "Godaaam! Wouldn't ya know it! Kenny! (the construction clerk) Get the rain gear out of the storeroom and get your ass out there and give the suits to the men."

Pete came charging in:

"Whadda ya wanna do, Earl?"

"Go ahead with the pour!"

Pete, with a toothy smile:

"I already told 'em that."

The rain intensified. It pounded. It rushed. It flooded.

Tse Tse came in, hat cocked to one side, hair soaked and clinging to his head, water dripping on his nose, his coat one big concrete splat. He was vocalizing, "Singing in the Rain." It splintered the tension and we all had a good laugh.

Then, Harvey, the resident engineer, showed up. He asked, "What are you going to do?"

Earl replied, "The men have been told to finish the pour."

Harvey, raising his voice:

"The concrete will be pockmarked. I'm not going to accept that kind of finish."

"That's up to you."

"Everything you do from now on is rejected."

"Put it in writing."

"If I were you, I'd call off the pour."

"Up yours."

It rained all afternoon and into the night. Pete and his finishers stayed until eight to completely trowel the concrete. Harvey went home at six. His inspectors stayed.

The next morning the slab looked diseased. But thousands of pounds of flash patch cement were purchased and the slab was troweled as smooth as polished wax. When we walked on it, our legs were reflected on the shining surface. Later, the slab was covered with dirt and dust from the saws of carpenters cutting wall forms. Still later, the floor was coated with a black asphalt waterproofing. Finally, it was under water as sewage passed through the equalization tank. Nothing was heard from Harvey about the finish. He knew it would eventually be covered but the specifications called for a smoothly troweled surface and that's what he insisted on.

A week after the pour, these men walked on the solid evidence of their labors. Construction workers get satisfaction from seeing the physical evidence of their work. I marveled that men, left alone to work and plan independently, can produce with intelligence and skill such creative results.

CONCRETE

Hands and face dappled gray with
The splatter of wet concrete that
Cracks fingers into bleeding claws.

All day you hold the trowel which
Fattens in your grip as you sweep
Back and forth, back and forth
Till the concrete stiffens.

The blood heavy in your neck,
Your tilted shoulders,
Your arms heavy as you hold down
Wet stones as the work holds you
Down all the hours of the day.

What you have done with your hands
Takes shape over a long time, slowly.
The work over,
We can stand on it.

2/About this case study: its genesis and research strategy

The thousand-yard pour illustrated many characteristics of construction work and workers. The rest of this study will document and systematically analyze these features. The main emphasis is on the culture of the workers in their work environment—the construction project. In this study, I will present data from four large construction projects researched over a four-year period. After presentation of the data, I will give the reader an overview of the construction industry which should reveal how construction worker behavior and values fit into the framework of the construction industry.

The ethnographic description of construction workers involves the following ideas:

1. The culture of construction workers is characterized by autonomy and control of the work process by skilled craftsmen.
2. Construction work proceeds through an informal, face-to-face process in which workers are aware of their own roles and the manner in which their individual work fits into the overall building process.
3. The technology of construction work is a handicraft technology in which workers provide their own hand tools and thereby maintain their own independence.
4. Construction work is dangerous and arduous, and the construction industry has the highest number of accidents, dangerous working conditions, and deaths than any other major industry in the United States.
5. Uncertainty is the norm in the construction industry, which is reflected in job insecurity for construction workers and a loose, informal, day-by-day type of administration on the construction job site.
6. Job satisfaction among construction workers is quite high as a result of worker autonomy, control over the work process, informal and loose supervision, high wages, work performed by integrated work groups, and pride in craft.
7. Construction workers exhibit features of an occupational community in their lifelong commitment to their crafts, the merging of their work and nonwork lives, and the acquisition of their self-images and identities from their work.

We shall see how these cultural features of construction workers relate to the technological and structural nature of the construction industry. Five main features are singled out: 1) uncertainty; 2) handicraft technology; 3) informal work administration; 4) the subcontracting system; and 5) the localized nature of the industry.

Construction is the largest single industry in the United States, employing more than 4 million workers. This study deals with construction workers as an occupational group. Occupational groups have been studied mainly by sociologists and some applied anthropologists. They are increasingly receiving attention from anthropologists, and the field will undoubtedly see more industrial ethnologies produced and published.

All my adult life, I have been in construction and have loved the excitement of building. Every new project is different and a new challenge. The design, shape, and technology of construction are always producing new concepts. And I have always loved the social as well as work contacts with construction men. Yet, I have also had a deep love of learning and academia. On a part-time basis, usually in evening adult-education programs, I have taught classes in various colleges for 13 years. Though I was paid for doing it, I would have taught for nothing. The exchange of ideas with students and colleagues was a contrast with the direct, I-believe-what-I-see-and-feel attitudes of construction, but it was satisfying to me in providing another type of experience. There is probably no greater satisfaction than seeing the physical evidence of your efforts—hence the joy of building, painting, sculpture, music. Perhaps academics have such a strong drive to publish because they too wish to see the concrete, physical embodiment of their mental efforts captured in something they can see and touch and feel.

While I was project manager on the Amherst sewage treatment plant project, some very dramatic events took place at work and I tried to write about them. One result of this effort was "The Thousand-Yard Pour," included in this book as Chapter 1. I also started to write poems about construction work and workers. I showed them to my coworkers and was encouraged by their response. It occurred to me that I was in the unique position of being an anthropologist employed where the action was. If I studied construction workers, I would need no artificial entry into some "community" or organization. I was already a member of the group.

I started searching for material that others had written about construction workers. I found almost nothing. There was plenty of technical literature about construction practices and how to manage contracting firms. There was an excellent book by Mike Cherry (1974) about high steel workers and a study by LeMasters (1975) done in a bar where construction workers regularly drank. There was also an unpublished doctoral thesis by Jeffrey Riemer (1975), an electrician who wrote about his experiences on a housing project. It was a good study, but it did not deal with the large majority of construction workers who are employed on the larger commercial, industrial, and heavy construction jobs in the country. I was basically on my own. This was good in a way, because I would have to assemble and organize my material before it would yield some system of organization and classification. I did find literature on other occupations, some of which were craft oriented, and this was helpful because of the possibilities for comparative data.

At first, I thought a chronological format would be the most interesting. I could do a series of ethnographic accounts dealing with different situations as they occurred on the various projects, with each incident or story illustrating a different aspect of the world of construction workers. I still think this would make the most readable and enjoyable account of construction work. However, it did not seem that

I could organize my material analytically in this format. This study is therefore organized by topics, and the ethnographic and anecdotal material is fitted into a classification of units of study which I believe best analyzes construction work. Perhaps my next project will be to let the construction workers speak for themselves and have the reader enjoy their stories, humor, and the drama of their lives as much as I have for more than 20 years.

One of the first problems a researcher faces in anthropological studies is to define his "unit of study." This is especially true if he is doing research on modern, industrialized societies. Faced with a bewildering variety of populations dispersed throughout metropolitan areas, the researcher has the difficult task of finding a bounded population that can be identified and subjected to focused investigation. In studying workers at their work place, there is a decided advantage:

> They are usually fixed in one work place from seven to eight hours each day, and as a result one can rely upon direct observation of human relations rather than trying to reconstruct what one would like to observe by reliance on the semantically booby-trapped and time-consuming analysis of interviews (Chapple 1958:830).

The basic method of this study is participant observation. Florence Kluckhohn (1940:331) described participant observation as ". . . conscious and systematic sharing, insofar as circumstances permit, in the life activities, and on occasion, in the interests and affects of a group of persons."

In his scientific role, the participant observer seeks to register, interpret, and conceptualize the social facts he observes. He is interested in people as they are, not as he thinks they ought to be. Through social involvement with his subjects, the researcher comes to understand and know their thoughts and values (see Bruyn 1966:15).

"CONSTRUCTION WORKERS" DEFINED

The term "construction workers" in this study refers to unionized construction men who build nonresidential structures—office buildings, schools, dams, bridges, sewage treatment plants, industrial plants, shopping centers, and public works such as roads, highways, and utilities. Seventy-five percent of the construction industry, according to its main weekly journal, *Engineering News-Record*, by employment and value of construction is engaged in nonresidential construction (April 14, 1977).

THE SETTING

This study took place in a metropolitan area that is predominantly industrial in its employment and economic characteristics. The metropolitan area has a large force of skilled and semiskilled workers in steel, auto, chemical, metalworking, rubber, and food processing. The area ranks eighteenth in the nation in dollar volume of output of industrial products (*Western New York*, July 1977). The workers in the study population were mainly Italian-American, Polish-American, Irish-American, and German-American, reflecting the ethnic composition of the urban region.

Data were gathered at four projects:

1. One project was a new library building for a state university. It was to house a library, the university administrative offices, a faculty government chamber, a theatre, two cafeterias, a student lounge, and a student affairs center. The building contained 400,000 square feet. At the height of its construction, 250 men were employed on a large site on which the huge building was located. Data on this project were gathered between April 1974 and May 1975. Cost of the project: $20,000,000.

2. Another undertaking was a sewage treatment plant designed for a population of 1 million people. The four sewage pumps had a combined capacity of 200 million gallons per day. The site on which the plant was located covered 30 acres and adjoined a creek into which the treated water was discharged. There were residential areas on three sides of the plant. Thirty-foot earth beams were built to screen the plant from the surrounding residents. The plant comprised 30 structures, including a pumping station, an incinerator plant, an oxygen-producing plant, and a number of rectangular and circular processing tanks. Data on this project were compiled between May 1975 and October 1976. Total cost of project: $130,000,000. Our contract for Phase I was $15,000,000.

3. A third project was another sewage treatment plant designed for a population of 1.5 million persons. It adjoined a river, which made the project site extremely cold in the winter. The river also created water problems for our excavation work. A feature of this plant was a sophisticated filtration system which disinfected the sewage discharge. This project involved the installation of sewer pipes under the river. Data for this project were collected between March 1976 and October 1976. Cost of the project: $24,000,000.

4. The fourth job, and the one on which I am presently working, is a road and bridge project in the downtown area of a western New York city. The project involves removing six miles of streets, installing new sewer and water lines, and repaving. It includes the installation of a new street lighting system, a new traffic light system for 38 intersections, 2000 signs of various functions and shapes, and 50,000 feet of curbs and sidewalks. The project also involves the building of four bridges, which span several streets and connect with an existing thruway. The data for this project were researched between August 1977 and November 1978. Cost of the project: $12,300,000.

STUDY POPULATION

The population for this study consists of all the workers present on the various projects outlined above, approximately 1000 construction men. All of them were union members, employed by the general contractor for whom I worked or by specialty trade subcontractors. They included carpenters, plumbers, steamfitters, operating engineers, bricklayers, masons, concrete finishers, electricians, painters, laborers, tile setters, drywall installers, insulation and asbestos workers, carpet installers, sheet metal men, ironworkers, roofers, glaziers, acoustical tile installers, telephone installers, kitchen equipment men, boilermakers, marble and granite setters, temperature control installers, surveyors, elevator constructors, and truck drivers. They also include architects, engineers, building inspectors, contractors, subcontractors, material suppliers, resident engineers, and future owners.

The construction workers studied, the types of unions, the skill requirements for various jobs, the types of contractors, and the types of projects are characteristic of the construction industry nationally. There are other factors which make this study population representative:

1. The metropolitan area in which they worked is a large urban area with a population of 1,300,000. Most nonresidential construction in the United States takes place in such urban areas.
2. The region in which the research was conducted is subject to codes which ensure that buildings constructed in metropolitan areas of this size will be similar in workmanship and materials used.
3. All buildings and roads must adhere to regulations of health and safety set forth by state and federal safety agencies. This adds to the representativeness of this study.

BASIC METHODOLOGY

The study method used was participant observation. In this technique, the investigator participates as a member of the group under study. His observations are made under the same conditions facing the members in their daily life activities and settings.

In my research, I was a participant before I was an observer. I was a member of the construction team before I decided to research construction workers. I was therefore in a position to know construction workers as an insider. Being a socialized member in the occupation, I had a knowledge of the language and vernacular of the members of the group. It was therefore possible for me to discount any misinformation. As a part of the world of construction, I was able to conduct my research in a natural setting; my presence did not contaminate the research environment.

One of the problems with being an insider is to avoid letting one's bias affect the research. As a socialized member of the group, there is a danger of stressing the positive and overlooking the negative. I tried to control for this in two ways. I was a member of an urban research group where I could have my material subject to review and to members' critiques. Secondly, I did not use my own views as representative of those of the construction workers I researched.

3/The construction worker

Construction workers are mainly highly paid skilled craftsmen. They enjoy good working conditions and occupy a top-level position among American blue-collar workers. The physical and social conditions of their work give rise to an identifiable occupational culture which stresses particularity, directness, and concreteness. It views the fruits of manual labor as more worthy than intellectual labor. It is a milieu in which human beings are placed before machines and social solidarity before occupational rank.

This chapter presents a general picture of construction workers comparable to the overview of the construction industry in Chapter 8.

BACKGROUND AND ENTRY INTO THE INDUSTRY

A large percentage of the workers studied in this study were Italian-Americans, who entered the construction industry by following their fathers and grandfathers who were prominent in the building industry.

> In the cities . . . the Italian immigrant became a common laborer. In New York he replaced the Irish and the Poles on the work gangs, building streets, skyscrapers and subways. . . . Everywhere in the states of the Northeast, starting at the turn of the century, he (the Italian) contributed the lion's share of toil to the building of roads, dams, canals, tunnels, bridges and subways (Lopreato 1970: 143–144).

Many of the construction workers observed were from working-class neighborhoods which contained other families in the construction industry. Forty out of 67 men questioned had relatives working in construction. Family and ethnic affiliation are important elements in bringing men into construction. A man's relative can get him into the union and into a job with a contractor. Some start as truck drivers. Others learn by working during summer vacations. On our downtown road project, we had more than ten young men from high school and college during the summer of 1978. Most worked as helpers or laborers. Without exception they were good workers, came to work on time, and said they would like to come back next year. Six had relatives on the job.

Seidman's profile of plumbers is similar to my findings on construction workers in general (1958:50):

> With few exceptions the plumbers had been reared in an urban environment; about a fourth of them (thirty-four rank-and-file plumbers were interviewed) had followed their fathers' trade and many others had plumbers among their relatives. . . . There were some among the plumbers who find the occupation a family tradition, followed it with as little conscious choice as the young man in a mining community gives to his entrance into the mines. As one plumber said: "My whole family are plumbers and I just followed along. . . . my oldest brother was in it at the time and I don't know why, but I had a craving for it."

George Strauss states that among the trade unions he investigated, in construction there was a period when "the union was only accepting sons of members" (Strauss 1958:95). The traditional method of becoming a union member and thereby gaining entry into construction is through *apprenticeship training.* Apprenticeship is the most widespread of the "formal" methods of getting into construction. More men enter by "informal" means, provided they can show proof of qualification. This proof ranges from answering questions put to him by the business agent to appearing before a union examining board. Sometimes a man has to put in a few hours of work on a project to the satisfaction of the foreman and steward. In general, business agents have considerable power in the matter. There are advantages to informal recruiting for the union: 1) the necessity of supervising apprentices is avoided; and 2) the apprenticeship program requires the union to predict manpower needs three to five years in advance. It might turn out there is not enough work for their union members in that period. Direct admission means men are taken in only when needed and when all regular members are working.

During the past three decades, training methods have substantially improved in the construction industry. Before that, apprenticeship involved little more than informal on-the-job training. Modern apprenticeship training involves a systematic program of instruction, including the following elements:

1. A formal course of classroom instruction.
2. On-the-job training.
3. Related instruction off the job site.
4. Satisfying a joint labor–management committee responsible for the apprentice's progress.

The International Labor Organization feels that the apprenticeship type of program is the best mechanism for imparting construction skills (ILO 1969:84).

In 1937, the National Apprenticeship Act established a federal apprenticeship policy in the United States (U.S. Department of Labor 1969). The Bureau of Apprenticeship and Training of the Department of Labor was created to administer this act. Minimum standards were established, including: 1) a minimum starting age; 2) a schedule of work processes to be learned; 3) the number of hours per year of organized instruction in technical subjects; 4) a wage schedule; 5) the method of supervision of the program; 6) evaluation of the apprentice's progress;

7) role of employers and employees; and 8) equal opportunity without regard to race, creed, or national origin. State Apprenticeship Councils were formed throughout the nation.

The routine administration of individual apprenticeship programs is by local joint apprenticeship committees (JACs), composed of an equal number of representatives from union and management. The local JACs set up specific standards for their programs and direct them. The apprenticeship programs are usually financed by assessing employers a few cents per hour for the manhours of labor employed.

Selection of apprentices is made by the JAC. An applicant must meet minimum standards, which vary from state to state and craft to craft. Most building trades require a high school education. In addition, there are residence and U.S. citizenship requirements in many crafts. There are minimum and maximum ages for starting apprentices, ranging from 17 to 30, with some exceptions made for veterans. Since the enactment of the Federal Civil Rights Act, an applicant is frequently given an aptitude test on which he has to achieve a minimum score. The applicant who meets the minimum qualifications is then interviewed by the JAC and ranked according to criteria the JAC considers relevant. After selection, apprentices enter a program, typically from three to five years, which includes on-the-job training and related class instruction. Starting pay is usually half the journeyman's wage and increases over the term of the apprenticeship to the full journeyman's wage. In some cases, the building trades unions and employers contract with the federal or state government to provide training to minority or disadvantaged youths. On our road project, we were required by contract to train 11 youths in various trades—ironworker, teamster, carpenter, operating engineer, and labor foreman.

The above description is an ideal outline of the way the apprenticeship program is supposed to work. For many unions, there seems to be few strong economic or political forces impelling them to do an adequate training job. Haber and Levinson state that employers are less than enthusiastic about apprenticeship training (1956:Chap. 6). When a business agent or employer takes a personal interest in a young man, then a sincere effort is made to give him training. One of our local construction union business agents told me he took an almost fatherly interest in all young men he placed in the apprenticeship program:

> I have to look after them. I just don't sponsor anyone. A lot of people ask me to get their sons or their friends in the union, but I take a long hard look at the boy before I say yes. Once I say yes I follow what they're doing like a parent.

Many times after an apprentice is hired, neither the union nor management takes the time or interest to ensure that the youth does not get "lost." I have heard apprentices complain that they are given "dirty" jobs the first week of training and stay on it for their entire apprenticeship term.

The construction industry has recruited its fair share of the nation's young manpower. Young men entering the construction field constituted 10.9 percent of all employed men 16 to 24 years of age in 1968 (Woldman 1968:No. 111). During the same year, 9.5 percent of all males employed in the United States were in construction (U.S. Bureau of the Census 1967).

A COMPARISON BETWEEN CONSTRUCTION WORKERS AND OTHER BLUE-COLLAR WORKERS

Working in the building trades is ultimately related to the social organization of the construction industry. Construction work tasks are as unique as the construction process is distinctive. Unlike the factory worker, a building craftsman's job is neither narrowly defined nor performed under close supervision. The repetitive nature of manufacturing work is absent. Instructions are general, not specific. For example, a typical day's work assignment to the crews on a construction project will sound something like the following:

> I want Wade on the street breaker on Swan Street. Let Bimbo and Chucky work with Marshall on the water line at Seneca. Put two trucks on slag and two on stone. If Ecol (sewer subcontractor) has no truck under his backhoe, let him cast the dirt in the street and we'll pick it up later. If we get any curbing in today, let Frenchie and Shorty unload it. After the brickies get done with the manhole at Sycamore, there's a catch basin pad to be poured at Genesee. Tell Casper to put the heaters in the hole and shore it and put tarps over it.

The men are given their work locations and operations and told what materials and equipment are available. The organization of the tasks and how the work is to be performed is left to the craftsmen and their foremen. The craftsman is responsible for the layout of his work and completing it within a certain time. As Stinchcombe remarked (1959:168–169):

> Administration in the construction industry depends upon a highly professionalized manual labor force . . . and . . . is more rational than bureaucratic administration in the face of economic and technical constraints on construction projects.

Here is an example of how it works:

I told a carpenter I needed some movable road signs and gave him a sketch of what I wanted. The signs had to meet certain state specifications for road building. I told the carpenter to figure out what materials he needed. After he gave me the list, I had the construction clerk order the material and ask for immediate delivery. Then I told the carpenter I needed the signs in two days. He said there was no way he could do it unless I gave him more men. I assigned two more carpenters to the job and the signs were ready when I needed them, built according to state department of transportation standards.

Practices of the trade regulate performance, pace, and quality of construction work. A journeyman must not only perform under loose supervision but must supervise himself. Construction contractors often have projects at widely separated sites. The maintenance of close supervision and continuous communication with each member of the work force is prohibitively expensive. Thus, the self-administration of the construction craftsman is the most efficient response to the peculiar circumstances of construction work. The journeyman is trained to perform organizational and planning functions that in other industries are the responsibilities of management.

In contrast to construction, in manufacturing both product and the work process are planned in advance by persons other than those in the work crew. Whereas the construction craftsman can see the total product and even plan what it will be, the mass production worker usually knows only one small part of a product, which thus has little significance for him. All of the planning is done for him.

Peter Drucker singles out three aspects of men at work, in contrasting the craftsman (of which the building tradesman is one example) with blue-collar workers in mass production.

Fitting the Man to the Job Drucker points out that a craftsman "performs a variety of processes in the course of his work. He constantly changes tools, rhythm, speed, and posture. He is bound to find at least one phase of the work congenial to him" (1950:169). In the mass production industry, on the other hand,

> each job embraces only a very small number of operations . . . the productivity of the mass production plant is geared, by and large, to the output of the wrongly placed worker. The "standard of performance" for any one operation is the amount of work that can be turned out by an "average person," that is, by someone without obvious physical or mental handicaps but also without any pronounced ability or liking for the job (1950:169).

Fitting the Job to the Man Whereas the craft worker is the master of his tools, using them to perform an entire range of complex operations, on the manufacturing assembly line there is "the tacit assumption that man is a badly designed, single-purpose machine tool . . . (Drucker 1950:172–173).

The Worker and the Work Team In construction, men work in teams, gangs, or pairs. Every craftsman has a helper. In construction, one must not only cooperate with one's immediate gang, but must coordinate with other trades. Each tradesman has a sense of the totality of a structure and the part that each trade plays in creating it. Therefore, all trades can share in the pride of creating something new and knowing the part that each played. In contrast, "mass production technology tends to isolate man from man. . . . The traditional automobile assembly line provides a visible example of this social isolation" (Drucker 1950:174). And among white-collar workers, "the business machine operations in a bank, insurance company, or mail-order house carry much further the split between the complete integration of the process and the complete isolation of the operator" (Drucker 1950:174).

Thus, work organized on a craft basis permits a worker to perform an entire complex of operations, gives him mastery over his tools, and provides a work environment that permits social integration and cooperation. Work organized on a mass production basis tends to give the worker a limited number of repetitive tasks, treats him as though he were himself a tool, and tends to isolate him in the work process from his work mates.

SOCIALIZATION

Occupational socialization is a process through which the recruit becomes a regular member of the group (Hughes 1971). Necessary skills and techniques are learned and are at least minimally internalized. Concomitantly, the recruit begins to identify himself with his occupation (Becker and Carper 1956). This formal enculturation process is completed when the initiate is fully accepted into the fraternity of regular members and at the same time identifies himself as a member (Hughes 1971).

The transformation process can be conceived of as a set of stages along a career line in which the new member is evaluated regularly and given increased responsibility and in wage increments as he progresses satisfactorily. The evaluations are typically based on two criteria—the explicit mastery of skills and techniques particular to the work and an acceptable adoption of a related set of implicit qualities that the new recruit is expected to display as a member of the group (Hughes 1971). In discussions of the training of apprentices, skills and techniques are usually the focus. But there are other tangible, yet less noticeable, aspects of the socialization process that take place in the transformation of a new recruit into a full-fledged member of the group. In this discussion I am going to focus on three—tools, costume, and jargon.

Tools The "tools of one's trade" are one of the most important accoutrements of the skilled craftsman. The tools themselves, how they are used, when they are used, and how they are taken care of, are important indicators of skill and expertise. The tools of a craftsman are not only essential for carrying out his work, but also reflect something of the worker himself. Much can be determined about a construction worker by the tools he has, how he displays them, and how he uses them. When a new craftsman arrives on a job, he is evaluated by the other workers prior to beginning work. The tools he brings with him serve as an accurate indicator of his ability and experience. Worn, quality tools reflect experience. Specialized tools, in addition to the required set of hand tools, reflect ability.

The display of tools is also an indicator of ability and experience. Many workers drive to work in their own pickup trucks. Some have tool boxes custom-built for their trucks which are attached to the side frame or fitted into the rear of vehicles. When a craftsman has a tool which is rarely owned by an individual and which is borrowed by others, his name becomes associated with the tool. It becomes "Dom's bolt cutter," or "Harold's banding tool" or "Chuckie's tapping machine." New apprentices often carry and display more tools than are necessary. The experienced journeyman carries only the essential tools. An experienced electrician knows it is not necessary to carry "all that iron" on his tool belt when a few tools—a six-foot folding rule, one channel-lock pliers, one screwdriver, and a pair of side cutters—are all that are needed for most jobs. Foremen, who pride themselves as directors and managers of the work being done, carry few, if any, tools. Usually they have only a six-foot folding rule, stuck in their back pocket.

Costume The adoption of specific pieces of work apparel serve as implicit indicators of the occupational socialization process. The hard hat, which is worn for safety reasons on major construction projects, very often serves an auxiliary purpose. Hard hats define status among the ranks of workers and, in some cases, facilitate occupational identification. The particular color and insignia on a hard hat, whether it is old or new, shiny or dull, can provide reliable information about the wearer. Workers from different trades typically wear different-colored hard hats. Newly hired workers are usually assigned a new, shiny hat. Superintendents and managers typically have unscuffed, shiny hard hats, with their names and positions on metallic tapes attached to the front of the hard hat. The seasoned craftsmen usually have old, scratched, and dull hard hats, reflecting many hours of hard work and which serve as status indicators. Some workers continue to wear their hard hats on their way home or to the local tavern.

Bib overalls are an additional part of many construction workers' costumes. They are usually worn over one's regular clothes and serve several functions. They are an insulator from inclement weather and from the dirt and dust of the job. Bib overalls also provide many pockets to carry tools and materials. Color is significant. Painters and plasterers usually wear white overalls, carpenters usually wear blue and white striped ones, and plumbers and electricians wear dark blue ones. Bibs are not very fashionable, and workers compound this by buying them a few sizes larger to accommodate tools. However, they are functional and most experienced craftsmen choose to wear them at one time or another.

Jargon Acquiring and using the jargon of one's trade so that it becomes second nature is another aspect of the socialization process. Each trade has its own particular way of referring to tools and materials which are used in the work process. Very often, these are shorthand terms that need no further elaboration, much like scientists, doctors, or lawyers use in their professional careers. Thus, plumbers talk about "nipples" (short pieces of pipe threaded on both ends) and carpenters refer to "two byes" (a piece of lumber supposed to be two inches thick by four inches wide, but is actually one and one half inches thick by three and one half inches wide). Masons refer to the mortar used to hold their bricks and blocks as "mud." This language sets off one craft from another, identifying those who use it as belonging to a particular group.

Not only does language become particular to the individual trade but association with certain projects leads to an argot that is part of the building site. On the library project, the central portion of the structure was called the "tower" by the men. This term was not used by the architect nor in any of the nomenclature associated with the project. The appellation was applied by the men because it was four stories higher than the east or west wings of the building. On the downtown Buffalo road project, the men who were performing the road work came to know what types of equipment were available on that particular project and referred to the equipment by various nicknames. Thus, one small bulldozer was known as "Gertrude." The machine used to break up the pavement was a "hog-knocker" and the demolition ball was called a "headache" ball.

Thus, becoming a construction worker involves more than simply learning the craft. It involves a process of looking, talking, and acting like a construction worker.

PERSONAL RELATIONSHIPS IN CONSTRUCTION

The construction industry is a major one which affects many other large industries. Yet, its principles of organization are unstandardized and informal. Personal relationships predominate rather than the impersonal rules and procedures that are so predominant among most large industries. Perhaps more than workers in other industries, construction workers function as independent units. Each man in seeking employment makes his own arrangements according to his personal contacts, preferences, and schedule.

Building projects are of relatively brief duration. Thus, a new labor force is assembled for each project. Craftsmen are constantly engaged in negotiations with employers or superintendents in their attempt to find desirable new employment as their current projects are finished. Employers, as they obtain new contracts, are continually seeking to assemble a work force to enable them to build their projects. Most of the negotiations involved in the hiring process are based on personal relationships. Construction workers in any locality rarely walk onto a project without being known previously by their employers. If they are not known to their employers, they are certainly known by other members of their trade.

The selection of the labor force for projects is almost entirely in the hands of foremen and superintendents. Once a superintendent is selected by management to supervise a project, the assembling of the work crews is handled by the designated super. Foremen and superintendents use personal standards with regard to the kind of men they wish to hire. A superintendent told me what type of men he intended to hire for a road project on which I served as project manager:

> I'm going to choose men who know the work but who are not prima donnas. I'm not interested in superstars. Maybe they'll give you 200 feet of pipe one day. The next day maybe some dust'll blow in their face and they'll take off on you. I'd rather have guys who are steady. They don't have to be geniuses. I don't want dummies either. But guys who are not afraid to throw on some rain gear when it's pouring or who'll get into a trench full of water or an operator who'll get down out of his cab and help push a piece of pipe home. Also, I want guys who'll come running when they're being held up and tell the truth. No cover-up guys. I don't want death-bed calls all the time from guys who should know how to handle an emergency. I want men who'll tell me when I'm screwing up but who'll listen when I holler.

Foremen and supers want to have harmonious crews on their jobs. They will not (nor are they expected to) accept choices of men made by the front office. There was an instance where the front office wanted us to rehire a man we had fired on the sewage treatment plant project. The superintendent's first reaction was negative. He would absolutely not take him back. One of the principals of the company came to our field shanty to argue for the man. The superintendent said the man was so disliked, "he had been punched more times than Muhammad Ali." The employer was in the position of imploring one of his own employees to hire another former employee. The superintendent finally relented but insisted that the rehired man take a cut in pay. The superintendent's wishes were followed by management, as is the case with all matters of hiring and firing.

Since foremen and superintendents do not have to accept choices made by the front office, they often put into effect their own biases when they hire. Ethnic background is a large factor in their choices. It is no accident that the company I work for has mostly Italian-American workers, since the management is Italian-American. The religious factor is related to the ethnic one—most of the men are Catholic (Italian, Irish, and Polish).

Despite the racial prejudices of foremen and superintendents, they must hire minority workers since federal contracts require it. Construction workers resent the requirement to hire minorities, especially when some of their friends or relatives are out of work. Once minority workers are employed, their particular skills and qualities become known, and if they are competent in their skill they are preferred over nonminorities. We have a blacktop foreman who is black. His prestige is so high as far as his skills are concerned that some of the white workers kiddingly but proudly exclaim when they feel they've done a good job, "I'm a white Willie," using the black foreman's name.

Since construction workers must be acceptable to superintendents and foremen, the latter have a great deal of power. Still, a skilled, competent tradesman is not afraid for his job and shows considerable independence toward his supervisors. Foremen and employers are frequently in the position of petitioning to try to induce a good craftsman to work for them. Competent and cooperative crews which produce increase the standing of those who hired them.

It is not unusual for a super or foreman, address book in hand, to drive around at night or on the weekend, contacting workers in their homes, trying to round up crews for a job that has to start quickly. This personal negotiation, while not always representing a balance of social and economic power, does involve mutual concessions and serves to enhance a construction man's sense of self-esteem.

The personalized system of work relationships is modified to some extent by the union hiring hall. However, the hiring hall is mainly for men either not well-known or not competent. A laborer's local business agent told me the following:

> We've got 2300 men in the local, with about 1600 working. That's a pretty high percentage for us for men with steady jobs. The other 700 are what we call floaters. We use them to fill in short-term jobs. Most are "just laborers." A laborer is not just a pick-and-shovel man the way he used to be. He has to know how to use tools; he has to run equipment sometimes. If a man can use the laser* or do a little welding he's more useful to the contractor. At the hall, when we call for a certain job, the more a man knows the more he can put up his hand. We're getting a younger bunch these days, more educated; they can do more than laborers used to do. The unskilled worker is no longer around in construction. Everyone has to be skilled to some extent. Otherwise, they're floaters and maybe get a flagman job or a "gofor" (errand boy) job, but they don't work steady.

As a worker and a member of a construction union, the average foreman does not want to hold himself aloof from his fellow craftsmen or exercise autocratic

* Laser beams are used in the laying of pipe. There are two parts, the instrument which projects a beam of light and the target. The laser helps to lay the pipe straight.

power. Skilled craftsmen in construction suffer little from shyness or apprehension in dealing with their supervisors. Many are friends or drinking buddies with their foremen. Construction workers do not view foremen as instruments of all-powerful corporations, as might be the case in the auto or chemical industries. In many cases, construction workers know their unions are more powerful than their employers. Dealing with supervision on a personal basis permits a construction worker to maintain his sense of independence and self-respect, so he has no feeling of anonymity.

Each foreman or superintendent tends to set up a circle of friends or relatives who become a labor pool from which he hires. Tests for hiring are personal ones. In almost every case there is no company policy to which a foreman must subscribe. As previously noted, ethnic background and religion are important. There are also personality attributes the foreman considers important. I questioned two superintendents about what they thought were important when they hired, and they cited the following:

Availability Supers need to schedule work and therefore rely on a man's being available when he is needed. We were blacktopping in December and had to make a decision each morning on the day we blacktopped (rather than the night before) whether or not we would work. There were certain men we could call on the morning we decided to work and they would get dressed and come to the job. We gave those men preference in hiring men for our projects.

Craft Skills A foreman may like a man, but if he is not competent he will not be hired.

Sociability Construction work is based on teams and gangs, and without sociability there is often no cooperation. Dissension can wreck a project, so men who foment dissension are avoided.

Good Judgment A man must have the ability to solve problems or if he cannot do so, to admit it. Covering up or trying to lie one's way out of a jam is bad judgment.

Initiative Construction workers are expected to perform independently. This often requires initiative in rounding up materials and equipment and figuring out a crew's next move.

A journeyman who expects to work steadily must make himself known to as wide a circle of foremen and superintendents as he can. To this end, he finds it advantageous and often pleasurable to be in contact with other construction workers after working hours. He knows the favorite bars or haunts of men in his trade. He will frequent these places after work to drink, shoot pool, play cards, and relax with his work mates. If his foreman or superintendent lives in his neighborhood, he may drop in to talk with him about the job or the company or just "shoot the shit." If a journeyman is out of work, he will visit jobs where he has friends or knows the super. Even if they have no work, he will ask the super or his friends to put in a good word for him with others who may be hiring. He does the same thing with the business agent at the union hiring hall. The result of this activity is that he will be on familiar and personal terms with those who will be hiring.

The relationship between the tradesman and his foreman or superintendent involves prestige considerations. The foreman must have regard for the individual tradesman's position. If a man has formerly been a foreman himself, he is treated accordingly. During our road project, the company asked us to use a man who had been a superintendent on another job which had finished. We already had all the supers and foremen we needed, and we had no choice but to assign the man to a labor crew. It was embarrassing for me to do it, and the former super sensed it. He reassured me that it was all right, that when the job got moving we might give him other tasks to perform. Many times a super has to convince a good man to subordinate himself for a while until an adjustment can be made in his position.

The fluid nature of construction work, with its constant change, requires adaptability on the part of construction workers. In many cases, a man does not know who may be working next to him until he arrives on the job. On large-scale projects, where there are larger numbers, there is a greater chance men working together will not have had previous contact. This requires some difficult adjustment. If a man cannot adjust he will be laid off. Foremen and superintendents do not judge the effectiveness of a journeyman solely on his output. He must also cooperate with others. Experienced construction workers know they must adjust their behavior to fit with other members of their own and other trades.

Bennie Graves summarizes the personalized recruiting of construction work in this quotation from his article on pipeline workers (1974:415):

> More than other technically advanced and rationalized industries, construction work seems to depend upon informal systems of personal relations to locate skilled workers where they are needed. Consequently, the purely technical requirements of a job may be markedly modified by particularistic criteria. Mohawks, for example, are prominent in the high steel construction of the northeastern United States; various ethnic groups dominate particular building crafts in Detroit; the pipeline construction industry seems to depend largely upon groups of friends and kinsmen to recruit its work crews.

Impersonality is not the pattern of relationships in construction. Job recruitment is done on the basis of personal relationships, with friends and kin favored over others (Myers 1946; Graves 1970). Learning of skills still involves father–son relationships to a significant degree (Strauss 1971:103). Work crews are usually familiar with each other and have worked together on several jobs. Foreman and journeymen have preferences with whom they wish to work, and a smart superintendent will take those preferences into consideration. Many construction firms are small enough for employers or supervisors to gather their work force for face-to-face meetings. Sometimes contractors are so small that employers work alongside their employees. Many employees have worked with their employers before the latter went into business.

Strong emotions, pride, and the force of personality are often contributing factors in the performance of construction work. For example, on a road project we had a masonry crew building manholes at the rate of 20 vertical feet per day. One January morning, Casper C., an assistant superintendent in charge of masonry work, challenged the men. He told them if they produced 30 feet they could leave when they finished. That day the men did 30 feet by 3:00 P.M., but did not go home and worked until 4:30, quitting time. When Kenny E., the foreman, came into the field office, he shouted:

"Don't expect 30 feet every day, Casper. We're not animals, goddam it."
Casper: "How come you did 30 feet? You do it when you feel like it? Is that it?"
Don N: (another bricklayer) "Bullshit! We can't break our ass like that every day. If conditions are right, we can produce."
Casper: "What do you mean, 'if conditions are right'?"
Kenny: "It's winter time. Those enclosures you're building—they suck! They don't protect anything!"
Casper: "What the fuck you want, a cathedral?"
Don: "No, but it's supposed to be above freezing in the hole. It does no good to give us heaters if the tarps are wide open and the wind is whipping our ass. The mortar is freezing on the blocks."
Casper: "But you still got 30 feet. What gives?"
Don: "Up your ass, Casper. Don't expect 30 feet every day. You know what a 12-inch solid block weighs?"
Casper: "My ass bleeds for you."
Don: "You try humping a hundred pounds all day."
Casper: "I call those blocks my birth control blocks. You lay them all day and when you get home you're no good for nothing, except to sack out and sleep."

After that confrontation, the men continued to produce 30 feet, complaining and cursing the evil ways of Casper. The force of Casper's personality, the men's pride, and the emotion of the challenge were all human, personal factors that led to an increase in output. Meanwhile, the men laying up the block developed a closeness with each other, as well as an admiration and respect for Casper.

Personal attachment and interdependence is probably highest among high steel workers. With death the penalty for mistakes, ironworkers on high steel are very particular about whom they work with. Working relations between members of a crew are so finely tuned that if someone is not present the entire crew may not work. Mike Cherry, a high steel worker, felt so attached to his partner he had dreams of death at the hands of a faceless partner (1974:132). Jack Haas describes the interdependence of high steel workers (1977:167–168):

> The processes of social control I have described—testing and controlling fellow workers and maintaining and enhancing individual and collective control over the work setting—are processes characteristic of occupations where danger is a perceived worker problem. The careful surveillance and testing of colleagues, particularly newcomers, the controlled actions belying any fear, and the unified efforts to increase worker autonomy are sociologically relevant outcomes in situations where workers face extreme danger.

PRIDE IN WORK

Most construction workers believe they work hard, contribute to society, and earn an "honest living." They feel they produce something real and tangible. They can see the physical evidence of what they accomplish, and the physical labor that tells on their bodies gives testimony to the integrity of their work.

Tom Wicker, a columnist for *The New York Times*, wrote that a major problem is the perceived lack of quality in contemporary American life (Sheehan 1975:191). Wicker cited, among other things, that people do not care about their work. Construction work has a moral code which extends back to the Middle Ages and which is associated with craftsmanship. It is an ideal which is often not achieved, but which remains as a norm in construction work. Every trade union agreement and every set of specifications has phrases like "good workmanship" and "according to good practices of the trade." It is not mere lip service. Craftsmen are proud of quality work and ashamed of shoddy work performance. Where hand and craft methods are important, as in construction, the traditional work ethic of workmanship and quality will continue to operate as a cultural ideal. Caplow expressed this notion in the following quote (1954:131):

> In the crafts, the most distinctive attitudes are those which revolve around contractual obligations, implicit or explicit, which specify the rights and duties of each in relation to all. These attitudes, which define a day's work, a fair wage, a good shop, a bad foreman, the right tools, a reasonable order, a one-man job, or the jurisdiction of a craft, are essentially moral. They contain a principle of justice drawn from two important sources: the medieval idea of the "just price" and the early modern theory that society is founded by a free contract.

In a study comparing the meaning of work for five occupational groups—steelworkers, coal miners, skilled craftsmen, salespeople, and physicians—"self-respect" was highest among the craft workers. In fact, it was twice as high for craft workers than for any of the other four occupations (Friedmann and Havighurst 1954:173). Self-respect is quite important among construction workers.

ELEMENTS OF SELF-RESPECT

Leadership This has nothing to do with formal leadership positions. There are men on construction projects who occupy formal leadership positions but command little respect because they are willing to manipulate and lie to protect their own personal situations. Those who enjoy their own self-respect and that of others exercise leadership for the benefit of others.

On one of our sewer plant projects, we were trying to install 84-inch pipe in a 25-foot trench. We lined the trench with steel sheets, held together by steel I-beam braces at the top to prevent the trench from collapsing inward. The pipe sections were 25 feet long and weighed 25,000 pounds. The cross braces were 12 feet apart, a distance necessitated by the earth pressure that had to be withstood on either side of the trench. In order to install the pipe, it would have to be tipped to be lowered into the trench and clear the braces. A wrong move would tip the crane lowering the pipe. The foreman and crane operator on the operation would not risk it. The resident engineer said it couldn't be done. The project superintendent took personal charge and responsibility for the work. He solved the problem by placing chokers (steel rope slings) on either end of the pipe attached to a hook and bar at the center of the pipe. This enabled him to tip the pipe without risking the danger of its slipping. The pipe had to be lowered carefully, and the superintendent personally directed the crane operator. After showing the foreman how it was done, he turned the operation over to him. That night, after work, the men were drinking at the local bar expressing admiration for the superintendent, who had been willing to run the risk of failure and criticism by taking personal charge of a difficult situation.

Knowledge of the Work A man who knows his trade exudes self-confidence and self-respect. He demonstrates his knowledge through work considered acceptable by his fellow tradesmen. Thus, he is treated as a full-fledged craftsman. The evaluation of a craftsman's work by his work mates and other trades gives the building tradesman his self-respect. Construction projects are work environments where men work among equals—those who have mastered the tools, materials, and techniques of their trades. If a man has done that, it would be most unusual for a

Figure 5. The "effluent" line at the Amherst treatment plant. Effluent is sewage on the discharge side of various structures. The sewage has now gone through all the necessary chemical treatment, with the exception of the chlorine tank, before it is discharged into Tonawanda Creek, pure enough to drink. The soil, called "gumbo" had no shear strength. You dig into it and 24 hours later the trench collapses, if steel sheeting is not used. (Photo courtesy of John Kaiser, Roy Crogan & Son.)

superintendent, foreman, or any other journeyman to treat that tradesman with anything less than respect. The journeyman would have to commit some flagrant act breaking the norms of workmanship or behavior.

Since knowledge and skill in construction are acquired through experience, older workers are treated with respect. A man with many years in the field is assumed to have acquired many "tricks of the trade." Rather than being looked upon as inadequate because of advanced age, as is the case in many other occupations, an older worker is treated as an "elder statesman." If his physical strength wanes, an older worker will be given a job suitable to his advanced years with no lessening of his own self-respect or the respect accorded him by others.

On the library project, we had a bricklayer foreman who had not finished high school but whose knowledge of his trade was known and highly respected. He could look at a wall and figure out how many courses of brick a wall required vertically and how many bricks there were in a row. He was a mathematical wizard in knowing the dimensions of the various types of bricks and blocks and how they could fit around windows, doors, corners, or curved or slanted shapes on a building. Other bricklayers would say that if he couldn't figure out a problem it was insoluble. He would look at a blueprint and comment, "Look, I'm just a dumb bricklayer and never finished high school, but there's something screwed up here." Then, a group of us would gather around him, and while we were wrestling with the problem, the man would come up with the solution—to our great relief, because, "if Tony couldn't figure it out, no one could."

Earning a Living Most people in American culture associate earning a good living with respect. At a party that the men on the library job threw for one of the labor foremen, retiring after 20 years, a construction laborer said to me:

> My job has given me the kind of income without which I wouldn't respect myself. Look at Iggy, just a laborer and he has enough to retire. And he gets a party to boot. I hope I'll be treated the same way when I retire.

Respect depends upon a tradesman's ability to "command the rate"—to get the full hourly rate for his craft. This is a broad, external indicator of one's position.

There is the knowledge that occupational hazards or job delays for which management is responsible may strike workers indiscriminately. Luck plays a large part in a construction man's calculations. If jobs follow one another regularly; if a job of unusual duration is secured; if contacts with a superintendent produce work at the proper time; if all or part of these conditions are present, then a man will have a good year. But who can tell about next year?

Physical Strength and Stamina These play a large part in determining construction workers' self-respect. It is associated with their ideas about maleness and manliness. But it is also associated with the requirements of the job. Much construction work involves hard physical labor under trying conditions. Construction men must develop the stamina to persevere through very adverse conditions—extreme cold; arm-weary shoveling; leg-weary sloshing through mud; the chilling effect of high winds; the backstraining pushing and lifting of heavy weights. Men who do this work are proud of their physical capabilities and admire those capable of great physical feats. That is why arm-wrestling, fights, football, weight lifting, and other physical sports are so popular among construction workers.

SPORTS, HORSEPLAY, KIDDING, AND STORYTELLING

Construction men like physical sports. They also like horseshoe pitching, bowling, and horseplay. On one of the sewer plant projects, during the summer months, every day at lunch time the men pitched horseshoes. Some of the matches were quite spirited, especially since the men like to place bets on some of them. Winter is bowling time for many construction workers. A high percentage either belong to leagues or bowl with friends at bowling alleys with bars where they can drink and bowl at the same time. Often, bowling night is one of their nights out with friends and work mates.

Construction workers are often aggressive and physical. Many enjoy physical horseplay. At one sewer plant project, one favorite pastime was ripping the clothes of one of the concrete finishers. Pete's clothing inevitably had some tear or hole. The men always teased him about his tattered work clothing. A number of times, one of the foremen and Pete would engage in a clothes-tearing fight. Each man would try to get his finger into a hole in the other man's shirt or pants and then pull. Away would come a shirt or pants leg, revealing bare limbs that would set off peals of laughter from the other men.

Some of the men liked to play the Italian game of "fingers" or "boom." Two men play it. The idea is to show a certain number of fingers and shout a number from two to ten. The one who calls the number that corresponds to the fingers shown wins a point. The game is played to five or ten points. The men shout and scream into each other's faces, flinging their arms and hands with energy and gusto, leering at each other with contemptuous disdain to show what they think of their opponents in mock contempt. Many times challenges are made between rivals, who bet a keg of beer or as much as $100 on a game. When the game is over, the loser slinks away in mock humiliation.

Kidding and comic stunts are a regular occurrence on the projects. On one job, we had a short-statured laborer called "Shortie." The men had the carpenter make a wooden platform in the shape of buttocks. Then they outlined a place on the lunch bench and labeled it "Reserved for Shortie's High Chair." On one of the sewer plant projects, several men placed a chair on a wooden platform for the concrete foreman. They drew several stars on it to make it look like a movie director's chair and labeled it, "The Chief."

Tse Tse, the "bucket man" on the thousand-yard pour, never wants for ideas for horseplay. One day, he tore a picture of a nude woman from *Playboy* and wired it to the handle of the concrete bucket before signaling the crane operator to hoist the bucket to a group of waiting men on a wall pour. A group of us watched him, and we waited expectantly to see the reaction of the men when they saw the picture. When the bucket was poured and lifted away, one of the men had the picture in his mouth, setting off howls of laughter.

On one sewer plant project, the men enjoyed telling sex stories and jokes at lunch time. A favorite was sex episodes. One man in particular had a reputation as a "cunt man." He enjoyed giving elaborate, detailed descriptions of his sexual activities, amid shouts of "Pervert" or "You're fulla shit" from his audience. The more his veracity was questioned, the more coarse his language and the more cynical his smile. He liked to boast of how he tricked his wife to leave the house so he could carry out his exploits. I don't believe anyone really knew how true his tales were.

Construction workers enjoy feasts. Feeds were a regular feature on one of the sewer plant projects. Many construction workers enjoy cooking. I recall one day on the library job when two men were working on the sixth floor, out on a scaffold at the very edge of the roof cornice. They were having an animated conversation. When I came within earshot, I heard them exchanging cooking recipes. On one feed we had during the sewer plant project, the men used concrete blocks and an oxyacetylene torch to make a stove. Several men brought in their favorite dishes and during lunch time, instead of the usual sandwiches, we all had a hot, gourmet meal. Another time, the men organized a "wild meat" feed, which featured only meat from animals that had been hunted. The meal was accompanied by hunting stories full of frustration, comic situations, and triumphs. This activity—the kidding and horseplay, the storytelling and feeds—contributes to group cohesion and produces in the men feelings of fellowship and affinity.

Among longshoremen, who share many characteristics with construction workers, Pilcher observed this about their kidding and horseplay (1972:102):

> The insults exchanged are such that in any other context they would express and elicit hostility, but which within the appropriate social contexts express and elicit not hostility or conflict, but rather personal friendliness and solidarity.

ATTITUDES TOWARD THE UNION

The loyalty given the union by construction workers reflects their understanding of benefits they receive. The union has raised wages, increased employment, reduced competition, and somewhat lessened job insecurity. It gives the construction worker social contacts with fellow craftsmen. The union keeps the building tradesman informed of trends in the industry. For all these reasons, the union occupies a more central place in the life of construction workers than might be the case with other groups of workers.

Construction workers view a good union official as one able to settle difficulties without strikes. Whatever their opinion of strike action, most construction workers would agree with the statement that I heard an ironworker express: "If the union tells me to go on strike, I'll have to go." Most men accept the union's power to call a strike and make it effective. In the region where the research for this study was conducted, the men attributed the absence of major strikes to a better and more educated class of leadership.

However, union officials are often criticized for giving special favors to friends or kin and spending the union's money too freely. Members resent high salaries and cars for business agents. Many see their union officers as politicians first, whose main concern is to try to stay in office. If a union official does something for the members, it is often viewed as trying to build up a following in order to maintain his position. In general, there is loyalty to the union rather than for any union officer or official. Even those who have relatives who are union officials speak of them as office holders rather than leaders. The real leaders, in the eyes of the men, are on the job solving construction problems.

Most construction workers exhibit fear of violating union rules regarding strikes and picket lines. During the library job, in the summer of 1975, several of the trades on the job anticipated the renewal of their contracts. All settled, except the ironworkers, who set up a picket line. It came at the height of the season, and many of the other trades were angry because they would lose work when it was available. The picket lines were set up and the men dared not cross it. One laborer commented:

> When people are on a picket line, they're trying to better their conditions, and it's not up to me to spoil it. I don't ask questions. They're construction guys like me and the union says I gotta respect their line, so I do it.

Another man, an operating engineer, said:

> We're not allowed to go through a picket line even if we wanted to. They'd be after us and our cards would be lifted. We'd have to go before that board and then . . . bam, I don't know what, but they don't let you off easy for a thing like that.

Thus, there is a mixture of attitudes: 1) a union man must accept the instructions of his officers, especially with regard to strikes and pickets; and 2) a union man must support the actions of fellow construction workers trying to better their conditions.

SENSE OF INSECURITY

One of the basic elements of the way of life of construction workers is their concern with stability of employment. The construction industry, comprising 5 to 6 percent of the gainfully employed, accounts for 12 percent of the unemployment in the United States (*Monthly Labor Review* 1971). The sequence of work on a construction project can interrupt a man's employment since he is needed only at certain times. In addition, there is the problem of weather. In the cold, rainy areas of the country a man can lose a substantial percentage of his income during winter months. Unemployment in construction during February is three times higher than in August (*Monthly Labor Review* 1971).

The loss of days of work in the construction field does not come in regular blocks of time. This often means that a man cannot fall back upon unemployment insurance since most states require some waiting period before benefits begin.

The pervasiveness of concern with the threat of unemployment manifests itself particularly when a job is in its completion state. When the sewer plant project was nearing its end, many of the men asked almost daily if we were bidding other jobs; if there were projects that were coming up that they did not know about; what my next project was going to be and if I needed men on it. The completion phase of a project is always a sad time because men who have worked together are now saying good-bye, and they feel the tension that results from ignorance of where the next project is or whether one will be working after the next pay day. Some of the men kid about losing their job to hide their concern. At the end of his book, *On High Steel*, Mike Cherry captures the mood of what it is like during the wind-down phase of a project (1974:204–205):

> Many more operations remained before our part of the job was done . . . minor alterations remained to be made in some of the still open cellar iron. All three derricks had to come down. Some welding was still to be done, and there were a few hundred odd bolts yet to be stuck. . . . There were a few bent beams that had to be heated and straightened.
>
> But this was all detail work, and dull. I stayed on despite the boredom, because I didn't know how long it might be before I could find another job, but the place had become strange. The shanty was practically empty. The nails that had held so many jackets and overalls and sweatshirts were nearly all unused. Patrick had gone when the raising gang was laid off, and Coley quit the following week. The punks (apprentices) were gone, and twice a day we argued, the eight of us remaining, about who would go down for the coffee.
>
> On June 24th, Crockett came over to me while I was making up a bundle of old cables and told me that it was my last day. He had to lay somebody off, he said, and the other stay-behinds were all local books. He sounded almost apologetic. He said something about being sorry that there wasn't anything he could do for me, and how it was too bad that I didn't have a local book. Then he handed me my check. This slowdown was temporary, he said, and perhaps we'd all get together again when things picked up. I told him it sure as hell wasn't his fault I didn't have the book and that I couldn't see where I had any complaints coming. I'd been employed in the city without a significant break since the first day I'd come down, local book or no, and had made pretty good money during that time. More money than enemies, anyway.

Construction men have a sense of their self-reliance and independence. They know the insecurities they must face in their industry, but those who are self-confident about their skills and their integrity know they will be in demand and will be able to find work again after a layoff. They may have to go to the hiring hall or begin contacting friends and relations to make the necessary connections that will lead to another job. But they maintain the spirit of confidence, reported by Mike Cherry in his fine book about high steel workers (1974), embodied in the following remark,

> Hey, you know I was lookin' for a job when I found this one.

CONSTRUCTION ERRORS

One of the negative impressions about construction workers is that in the building process they commit many mistakes. There is a growing feeling among Americans that the quality of work in general has deteriorated and there is little pride in workmanship in today's America. Perhaps construction workers are particularly vulnerable to this charge because the public has direct contact with them, at least in the housing sector, where so many Americans own their own homes and have been involved in some construction work, alteration or new. The other element is that in construction, one is dealing with a bulky, complex product that takes a long time to produce, is affected by weather, and is subject to the work of many hands. Construction cannot be given the same type of environmental control that is attained in a factory, and therefore the acceptable tolerances are greater. Furthermore, the fact that so many different parties are involved with the planning, design, and layout of structures increases the chances of errors.

Mistakes are the inevitable result of any kind of work or activity. They are, at the very least, the unintended result of any type of human effort. The treatment of errors by construction workers is no different from the manner in which errors are treated in the society as a whole. Trade-offs are made whereby the seriousness of a mistake is weighed against its physical or human consequences. If an error could endanger persons or cause damage to property, there is no decision to be made—it must be corrected. Any other course of action would be legally criminal, if the defect were known and wilfully ignored.

In most of the other cases, in which less serious errors occur, contractors, architects, and construction workers weigh the seriousness of a mistake against the cost of correcting it. If the fall of a pipe was less than that specified but would still work and permit water within it to drain by gravity, it would be foolish to rip it out and reinstall it. Here is an illustration of a bureaucratic and then a practical approach to an error:

On the downtown Buffalo road project, we built four bridges and purchased steel from a company in Pennsylvania. The State Department of Transportation specifications stated that we had to apply a new shop coat (protective red lead paint applied in the factory to prevent the steel from rusting) if the steel sat outside for more than 120 days. Through our own errors of timing and coordination, the steel sat outside all winter. When the job was finally begun, the shop coat looked perfect, except for a few scratches. The state inspector and project engineer, going "strictly by the book," demanded that we repaint the steel, even though they admitted that it looked good. This would have cost thousands of dollars. We appealed to the state transportation department, and a state inspector from Albany was sent to examine the steel. In this case, reason prevailed because he examined the paint and said it was fine and we only had to touch up the scratches.

Construction workers are aware that their careers and the longevity of their employment depend on their performance. In the case of construction workers, this may be more true than in other occupations because performance in construction can easily be pinpointed to the crew and the man doing the work. Construction, with its hand technology and labor-intensive work process, depends on the individual worker for the quantity and quality of work. Thus, there is a strong reason that most construction workers desire to perform reasonably well. They will gain the respect of their fellow craftsmen, and a reputation for good work will further a man's future employment prospects. Construction workers who just put in time or do sloppy, careless work are not the first to be sought after when workers are being assembled for a job. Especially with the localized, community-like nature of the industry, workers are known by foremen and supers who are aware of the type of work they do.

For those who question the attitudes and responsibility toward work of construction workers, I wish to point out another aspect of construction workers' behavior with the following example. The company I work for was the contractor which did the clean-up work of the Love Canal, the biggest environmental disaster caused by chemical contamination in the United States. We built the first treatment plant to take care of "leachate" (chemically saturated water leaching from the canal into adjoining property). Our work included installation of a drainage system, underground holding tanks, and the new plant. We also capped the contaminated area of the Love Canal with three and a half feet of clean clay.

We had 28 men working on the project. They had to take showers every day after work. They had to wear special suits over their work clothes, which were discarded each day to be washed and decontaminated. The men had to use respirators whenever they worked in a manhole or in a deep excavation. They were working in soil often filled with water contaminated by one of the most deadly chemicals—dioxin. This chemical ate up steel pipe and even polypropolene pipe, which is a plastic pipe resistant to many corrosive materials. The men worked in an environment putrid with the odor of chemicals. They received no extra or premium pay for this work. They and the State Department of Environmental Protection had no idea of what they would encounter when they dug into the earth. Yet, not a single man quit or refused to perform all the work that was required of him or objected to using the various uncomfortable and clumsy safety devices while he was performing his tasks. The only incident that occurred was when a telephone installer came to put phones in our trailers. He took one sniff of the air, said, "This is not for me," and left. Our own men stayed on the job from start to finish, learned to use all the safety devices, and completed the project without any serious incidents.

When people talk about errors or mistakes in construction, they should also be aware of the dangers construction men face and overcome in order to give the general public the homes, highways, dams, bridges, office buildings, and beautiful monuments they enjoy. People should think about the high steel workers who risk their lives on 40- and 50-story skyscrapers, or painters who hang on bridges to paint them, or road workers who may inconvenience the public but who work in all kinds of weather so that people will have the ability to travel and deliver goods from one place to another.

Construction workers, as well as other blue-collar workers, should be given recognition for all the dirty, heavy, smelly, uncomfortable jobs that many people would not dare to do themselves. After all, blue-collar workers do not get a good press because the people who write about them are not blue-collar workers themselves, except in very rare cases. By and large, what working people do is largely ignored. For example, when Jane Fonda came to the Love Canal, she covered herself with publicity and glory as a fighter for saving our environment. She addressed a crowd of 400 people outside the gate, in front of the press and the TV cameras. However, when she entered the contaminated construction site, she sped through the area in her car with the windows closed, and she never took the time or trouble to stop and talk to the workers who were cleaning up the mess.

BIG RED
(*Dedicated to Kenny, the bricklayer*)

He locked dumb blocks into place
And a manhole rose from the ground,
He has earned good bread today.

The March wind wraps the street
In sharp shadows of falling dusk,
The job is a ghastly beast
Filled with gaping holes.

The day's brutal conflict forgiven,
He prostrates himself so swollen veins
Send yearning blood coursing through
That mountained body's leaping streams.

Tomorrow he will return, that huge hand
To knead the heavy mortar,
And the next day and the next.
As he said, his life would be
Left on brick piles.

4/The construction worker on the job site

CRAFT ROLES

One of the important roles on a construction project is the craft role. To perform a craft role, a man must possess certain skills and knowledge. He must display this knowledge and skill in what he produces and the services he performs on the project. The carpenter's knowledge will result in the building of wooden products; a mason's stone; an ironworker's steel. Each craft contributes to the total building scene. The tradition of the master builder who performs all tasks is no longer possible because of the size and complexity of structures. There are few individuals in our society still capable of building a house and performing all the tasks themselves. Each craft has its own fund of knowledge and skills and its particular functions and responsibilities.

There are 20 trades in construction, organized into unions, plus the teamsters. In Table 1, I have covered the main ones. Some of the trades not covered are specialty crafts, like elevator construction. Others are finishing trades like acoustic ceiling men and asbestos workers. Other crafts, like plasterers and marble polishers, are few in number and becoming rare in their use.

In the table, I have used the term *function* as a service that a craftsman renders by virtue of his special training. The term *responsibility* is used to denote an *obligation* to perform the craft service or to behave in a manner which conforms to craft or construction norms of behavior.

PRESTIGE

There are some factors in construction that are associated with prestige. Generally, the greater the skill, the larger the wage scale, the greater the danger, and the cleaner the work, the greater the prestige. For those crafts that require more skills than others—carpentry, stone cutting, masonry, electrical wiring, and plumbing—the more the craftsman will be permitted to organize his own time and lay out his own work. Discretionary use of one's time is highly valued and those who have it enjoy high prestige. For those crafts that are highly skilled, there are fewer men in other trades who are familiar with the skill. For example, most construction men will say that they make it a practice not to get involved with electricity because they don't know enough about it. There is a great deal of kidding about prestige. Men in the lower skilled trades frequently kid with the more skilled workers by offering to exchange paychecks with them.

The association of dirty work with less prestigious work is often expressed by foremen who threaten to "punish" a man by sending him into a "hole" or the "mud" or a "sewer." One of the laborers on the road project was known as "dirty Johnny" because of his appearance. He was usually given the dirtiest jobs. He identified himself with the low prestige that related to his appearance by trying to disgust his fellow workers with his descriptions of his experiences while doing sewer work. He enjoyed describing the size and shapes of feces or other objects he saw floating in a sewer pipe. One day we had to tie a live sewer line and needed a man to work inside the pipe. Johnny was selected and he jumped into the pipe with laughter and alacrity. When he came out, smelling awful, he said, "There'll be no flies in Buffalo tonight because they're all going home with me."

VALUES AND SANCTIONS

A successful project in construction benefits all those associated with it. The employer realizes his anticipated profit and feels positive about those who participated in the work. The men on the project know when things are going well. The two most important elements of a successful project are completing it on time and performing the work in accordance with the plans and specifications. The men on the project know if target dates are being met. They also know if the quality of the work is acceptable through their contact with the owner's inspectors. A positive atmosphere on a project is self-propelling in the good relationships it engenders between trades and between the owner and contractor. Men associated with the project are often rewarded with other work or are sought after when new work comes up. A bad project promotes all the negative behavior that goes with construction mistakes, losses, and a bad reputation for the contractor, the men, and the supervision.

Table 2 lists the values and sanctions that can result in a successful or unsuccessful project. The table on crafts dealt with knowledge and skills. This one deals with social behavior. Construction workers learn the norms and values of their trade as part of their socialization process, and they are as important as knowledge and skills.

THE DIMENSION OF TIME ON A CONSTRUCTION PROJECT

Every construction project has a limited time frame. It has been estimated that for a 10- to 20-million-dollar contract, it takes approximately ten years to conceive, plan, design, finance, and build it. Some projects stay on the shelf for a long time before they are activated. Our downtown road job was put off for 20

TABLE 1. CRAFT ROLES–EMIC

Craft	*Knowledge*	*Skills*	*Function*	*Responsibility*
1. Carpenter journeyman	a. Geometry b. Measurement c. Reading blueprints d. Properties of wood and metal substitute for wood	a. Hammering b. Sawing c. Fitting d. Fastening e. Glueing f. Measuring g. Leveling h. Shimming i. Plumbing	a. Frames b. Forms c. Finishes	a. Provides own tools b. Builds from blueprints or verbal orders c. Lays out own work d. Directs helper e. Directs apprentice
2. Mason journeyman	a. Geometry b. Measurement c. Reading blueprints d. Properties of brick, block, stone, and mortar	a. Laying b. Buttering (applying mortar) c. Leveling d. Bedding e. Striking f. Parging g. Breaking h. Sawing i. Measuring j. Counting k. Coursing l. Plumbing	a. Builds and faces walls b. Builds veneers c. Builds foundations	a. Provides own tools b. Builds from blueprints or verbal orders c. Lays out own work d. Directs helper e. Directs apprentice
3. Operating engineer	a. Workings of controls and devices on construction machines b. Load capacities	With mechanical means: a. Lifting b. Lowering c. Blading	a. Moves and places materials b. Cuts and borrows; excavates and embanks dirt	a. Operates construction machines safely b. Follows verbal orders c. Follows hand

Craft	Knowledge	Skills	Function	Responsibility
	of construction machines c. Safety procedures d. Properties of earth	d. Tracking e. Pulling f. Pushing g. Banging h. Digging i. Placing j. Leveling k. Hauling l. Loading m. Unloading n. Hammering o. Ripping p. Tamping q. Excavating r. Backfilling s. Clamming t. Tugging	c. Shapes earth d. Grades earth	signals d. Protects equipment
4. Ironworker	a. Measurement b. Blueprints c. Size and weights of steel members and rebars	a. Balancing b. Bolting c. Riveting d. Pushing e. Pulling f. Shimming g. Climbing h. Shimmying i. Sliding j. Cooning k. Torch cutting l. Welding m. Leveling n. Plumbing	a. Erects and connects structural steel b. Places and ties reinforcing steel c. Fabricates and fits miscellaneous iron d. Erects iron stairs	a. Furnishes own hand tools b. Erects steel true and plumb

TABLE 1. CRAFT ROLES—EMIC *(continued)*

Craft	*Knowledge*	*Skills*	*Function*	*Responsibility*
5. Laborer	a. Measurement b. Blueprint reading c. Oxyacetylene burning d. Use of laser beam e. Use of cement mortar f. Use of power tools g. Use of dynamite h. Use of underground pipe i. Use of concrete j. Use of jack hammer k. Use of bush hammer	a. Shoveling b. Carrying c. Mixing d. Placing e. Screeding f. Leveling g. Breaking h. Hammering i. Busting j. Burning k. Blasting l. Laying m. Loading n. Unloading	a. Handles all heavy hand labor b. Assists journeymen	a. Performs hand labor under foreman's orders b. Protects tools and materials c. Cleans up
6. Electrician	a. Electrical circuitry b. Overhead and underground power systems c. Electrical devices—switches, panels, fixtures, controllers d. Electric motors e. Lighting systems f. Blueprints g. Measurement	a. Piping b. Wire pulling c. Wire connecting d. Panel building e. Circuit tracing f. Testing g. Device mounting h. Connecting fixtures and motors i. Measuring j. Installing power manholes k. Grounding l. Energizing m. Interlocking	a. Installs electrical and power systems	a. Provides own tools b. Performs work according to Safety Code c. Energizes electrical equipment and devices d. Repairs and restores interruptions of electric service

Craft	*Knowledge*	*Skills*	*Function*	*Responsibility*
7. Plumbers and steamfitters	a. Pipe, fittings and valves b. Drainage and waste systems c. Water systems d. Heating systems e. Air conditioning systems f. Boilers g. Air conditioning plants h. Humidifiers i. Filters j. Plumbing fixtures k. Radiation l. Blueprints m. Measurement	a. Fitting b. Threading c. Cutting d. Welding e. Soldering f. Brazing g. Rigging h. Hanging i. Fastening j. Caulking k. Leading l. Testing m. Repairing leaks n. Pressurizing o. Bleeding p. Charging refrigerants	a. Install drainage, waste, water, heating, and air conditioning systems b. Install compressed air and medical gas systems c. Install industrial chemical and liquid systems d. Install waste water treatment systems	a. Install systems according to blueprints and specifications free of leaks and defects b. Repair, replace and and restore any defective installation c. Install and test all required safety devices
8. Painter and paperhanger	a. Properties of paint, removers, varnishes, shellac, wall paper, and other wall coatings and coverings	a. Brushing b. Pasting c. Glueing d. Matching e. Fitting f. Removing g. Sanding h. Mixing i. Blending	a. Apply finish coating and covering to walls	a. Mix paint colors b. Apply designated color to designated wall c. Match patterns d. Apply coating and covering neatly e. Protect uncoated surfaces

TABLE 1. CRAFT ROLES—EMIC *(continued)*

Craft	*Knowledge*	*Skills*	*Function*	*Responsibility*
9. Concrete finisher	a. Properties of concrete	a. Floating b. Troweling c. Screeding d. Pouring e. Vibrating f. Loading and unloading buckets g. Pumping h. Bullfloating i. Raking	a. Pours and finishes concrete true and level or pitched if required	a. Furnishes own tools b. Finish as specified c. Finish during proper temperature and moisture conditions
10. Sheet metal worker	a. Properties of sheet metal b. Blueprints c. Types of sheet metal locks d. Types of hangers e. Geometry f. Measurement g. Air flow mechanics h. Fan mechanics	a. Cutting b. Fitting c. Locking d. Connecting e. Banging f. Hanging g. Measuring	a. Installs and connects sheet metal duct system as part of heating and air conditioning systems b. Installs and connects fan and blower systems c. Installs and connects ventilation and exhaust systems	a. Furnishes own tools b. Makes connections without leaks c. Installs correct size and shape of ducts and air outlets
11. Teamster	a. Properties of trucks, lowboys, and concrete mixers	a. Driving b. Braking c. Turning d. Backing e. Dumping f. Operating controls g. Applying water h. Turning mixing drum i. Cleaning j. Washing k. Checking	a. Operates wheeled truck vehicles to deliver and sometimes unload materials for construction project b. Moves earth, dirt, or debris from one part of job site to another or off the job site	a. Has truck at the right place and at the right time b. Operates truck safely c. Protects equipment

TABLE 2. VALUES AND SANCTIONS

Values	*How Expressed in Behavior*	*Positive Results for Conforming*	*Negative Results for Not Conforming*
1. Getting to work on time	a. Arriving early and socializing before starting work	a. Companionship b. Fellowship c. Compatibility	a. Docked pay b. Ridicule c. Disrespect
2. Performing work correctly and efficiently	a. Performing work once without the necessity to redo it	a. Assignment to responsible or interesting tasks b. Promotion to foreman status	a. Complaints from foreman or super b. Layoff c. No rehiring for other projects
3. Cooperating with other crafts	a. Performing a work task at request of another craftsman or foreman b. Permitting other trade to complete its work before they are prevented from doing so by cooperating trade c. Performing work task so as to make tasks of other trades easier	a. Harmonious relationships b. Pooling of resources c. Mutual assistance	a. Quarrels and disputes b. Isolation c. Bad relations with other trades
4. Following instructions and orders	a. Doing exactly as instructed and in a timely fashion b. Making sure orders are understood	a. Fulfillment of needs of the project b. Fulfillment of the role assigned c. Performance of one's function d. Sense of achievement	a. Criticism for breach of trust b. Dismissal from job c. Sense of failure
5. Being honest about one's work	a. Admitting mistakes b. Taking responsibility for one's own errors c. Performing corrective work if necessary	a. Others' trust b. Good reputation	a. Others' suspicion b. Humiliation when caught in a lie c. Dismissal from job

Values	*How Expressed in Behavior*	*Positive Results for Conforming*	*Negative Results for Not Conforming*
6. Being willing to perform difficult or dangerous work	a. Exposing oneself to risk in an excavation, on a scaffold, or on a high perch like the edge of a building or structural steel b. Working in burdensome conditions like extreme cold or heat c. Working in dirty conditions —mud, sewers, tunnels d. Doing heavy, onerous, backbreaking, or grueling work like pushing concrete all day or lifting heavy loads	a. Respect and admiration from other men b. Expressed praise c. Selection for special tasks d. Selection for difficult tasks because of trust placed in the person	a. Contempt or ridicule (unless old or hurt)
7. Taking care of tools and equipment	a. Reporting malfunction in equipment promptly b. Picking up tools at the end of the workday c. Using tools with care and safety	a. Others' trust in using best equipment b. Reputation for safe operation c. Reputation for protecting others d. Consultation on types or makes of tools and equipment	a. Dismissal for abusing equipment b. Object of expressed anger of other men, most of whom particularly resent abuse of tools and equipment c. Dismissed or docked pay for leaving tools out where they can be stolen

years. A construction project begins after the bidding process leads to the selection of a low-bid contractor and the signing of a contract. The contractor is given a letter to proceed by a certain date. When all work is performed to the satisfaction of the architect or engineer representing the owner, the contract is considered complete and the owner makes final payment.

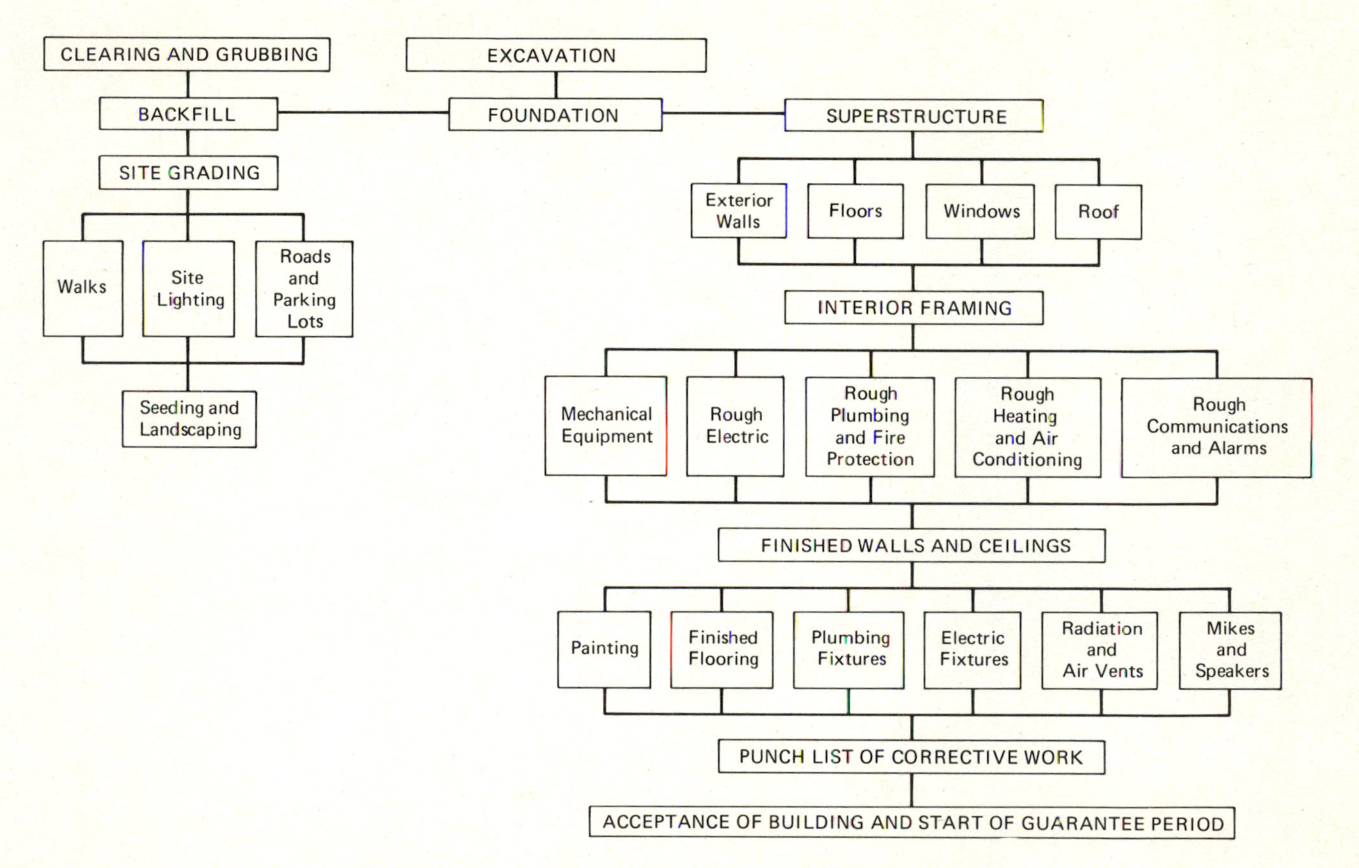

Figure 6. Key construction phases on building projects.

During the construction process, there are several crucial beginnings and endings affecting interlocking phases of the work. Various methods have been devised to assist the contractor in developing a systems approach to scheduling the different phases of construction work. One of these is called "Critical Path Method" (CPM). It is based on listing all events that will take place on the construction job and assigning time in hours or days for completing each phase. These events are then plotted in a network diagram with arrows outlining those events which logically precede others (see Figure 6). The critical path is the stream of events that takes the longest time.

In addition to the beginnings and ends of construction phases and the total start and finish of projects, there are the daily beginnings and ends. The workday starts early, 6:30 to 7:00 A.M., with foremen and superintendents arriving to plan for the day's activities. The men arrive from 5 to 30 minutes before the 8:00 A.M. starting time. They socialize, have coffee, and receive their assignments for the day. During the day, men might be shifted several times from task to task. The day's work is interrupted by a half-hour lunch break and a 15-minute coffee break in the morning and afternoon. Work ends at 4:30 P.M. Some men report before the regular starting time if a special task is required. On a road project, a street might have to be kept free of cars and thus a flagman will come in early and wave all cars away. In winter, mason helpers report early to heat the sand so it will be usable by 8:00 A.M. Some men work after 4:30 to clean up or to complete a work task. If rain is expected that night or the next day, an embankment might be rolled so the rain will run off rather than saturate the earth work.

REGULARLY RECURRING ACTIVITY

The essence of organization is a regular and recurring activity. Social organization consists of patterned recurrence of behavior which persists for varying lengths of time. The regularities and uniformities are structured into the behavior by rules and standards which we term norms. Some of the rules come from the union. For several crafts, union rules require that the first man hired must be the steward. Other rules come from external sources, such as public agencies which govern safety. The Office of Safety and Health Administration requires that any excavation deeper than five feet must be shored with wood or steel panels to protect the men working in the excavation.

While rules and standards function to regularize work behavior, regularization is somewhat mitigated by frequent changes. It is unusual for a man to be at a single work station for any length of time. A week would be a long period. He may perform the same activity, but he would do it in different locations and under varying conditions. In this example, building manholes on a road project, the following is the recurring pattern of activity:

1. Each morning two laborers in a four-man crew come to work a half hour early. If it is cold, they heat the sand. They load sand and mortar on their crew-assigned truck. They check the mortar mixer, make sure it starts, and attach it to the back of the truck. They put the mason's tools on the truck and are ready to pro-

duce by 8:00 A.M. The previous day they had placed concrete blocks and bricks at the manhole location where work was to be started or continued.

2. At 8:00 A.M., with the two bricklayers riding in the cab and the laborers in the back of the truck, they proceed to the work area.

3. When they arrive at the excavated area for the manhole, the bricklayers go down into the hole. Before they do, the laborers check the shoring and bracing, and they adjust and tighten them to make sure they are secure. If any mud has slid in the hole, all four men work to shovel it out. Bricklayers are not supposed to shovel according to union rules, but the informal norm on this job is that they do.

4. When the hole is ready, one laborer stays in the hole and the other goes on top. The man on top mixes the mortar and sand in the mixer and sends it down in pails. He also slides down blocks on a wooden chute. The laborer in the hole catches the blocks and hands them to the bricklayers. The 12-inch solid blocks weigh 100 pounds and have no opening at the top, making them difficult to handle. The bricklayer must cradle them in his arms before he sets them. The laborer in the hole places mortar boards where they are reachable for the bricklayer. The laborer piles the mortar board with fresh mortar as the bricklayers use up the mortar with which they bed and bond the blocks set in place.

5. The bricklayers install the blocks handed to them; they set them in a bed of mortar and then fill in the sides. They plumb and level each block as it is laid, using their eyes and hand levels. With the levels they check the top of the block for level and the side for plumb. When the manhole is completed to the top, it is "parged" on the outside—a two-inch layer of mortar is spread with a trowel to seal it to prevent water infiltration.

The above description is what is supposed to take place each day. However, something always happens to vary the pattern. If it rains, time is spent pumping water out of the excavation. If the surveyor has a "bust" (makes a mistake), the manhole might be in the wrong place and might have to be torn down and rebuilt. Some days the mortar mixer breaks down. Many times the banks of the excavation cave in so severely they cannot be dug out by hand and a machine must be used to redig it. Sometimes a gas, water, electric, or telephone line is in the way of the manhole and the crew must move to another location until the utility company cuts and moves their line. On some occasions, a manhole crew catches up to the sewer line crew and gets laid off for a week or two. Other times, they fall behind and a second crew must be hired. From time to time, someone gets hurt or there is an accident. A block might jump off the chute and injure the laborer, or the bricklayer might get some mortar in his eye.

With the many possibilities in construction for interruptions in regularized work activities, recurring activity is only maintained through solving the frequent, sometimes daily, breaks in the planned work process.

RECURRING SOCIAL SITUATIONS

The Physical Framework The construction site is a physical frame within which social events take place. The site is subdivided according to function and the designed placement of structures. Trailers and work shanties are located near work stations. They are places where men dress, keep their tools, eat lunch, and socialize. There are other areas where material compounds and storage trailers are kept. These areas are usually fenced in with a locked gate. Heavy machinery is also kept in a locked compound area which also contains a fuel pump and gasoline tank. There are locations for water supply and toilet facilities. There is always a public telephone on the job as well as phones in the various shanties and trailers.

Sometimes, the entire construction site is fenced in and guarded by watchmen. The site and its various facilities become familiar to the men who learn how to use them to satisfy their needs. They also learn where to find the persons they need to communicate with to perform their work or to have questions answered and problems solved. As the construction project proceeds, certain trades are associated with their own site location where they have their shanties, materials, and social activities.

Events Events are associated with work operations. These events have social consequences because they involve various individuals acting in work roles with each other. Work events involve communications between foreman and journeyman, among journeymen, and between journeymen of different trades. There is also a social network that involves inspectors, employers, owner's representatives, and union representatives. Failure to consider the social consequences of events can have adverse effects on a construction project. Other trades might have to be notified not to work in a location until certain other work is completed. Problems, mistakes, or information might have to be communicated to the architect for analysis and decision. If a craftsman is not sure of the intent of the plans and proceeds with the work event without engaging other responsible parties, he has ignored the social consequences and it could have far-reaching adverse effects. Such an act could risk collapsing a section of a building or costly removal and reinstallation of a major section of a structure.

Work events can have social consequences involving nonconstruction people. On a road project, when streets have to be closed or restricted, public agencies like the police and fire departments are involved, as well as ambulance companies, hospitals, and bus companies. If traffic has to be detoured, the driving public is inconvenienced. Newspapers, radio stations, and TV stations must be made aware of detours and closings so as to warn the public. On road projects, construction workers often become directly involved with the public who may physically be in the way, preventing the men from performing their work. Sometimes, the people act irrationally, such as when drivers attack flagmen trying to direct traffic around a congested work area.

The social consequences of work events are as predictable as the physical consequences of applying work skills. They must be considered in the planning for work events. Failure to do so can lead to drastic action such as the shutting down of an entire project by the police or the project owner. Social events associated with work events are as important in the execution of construction work as the applying of craft skills to the physical work itself.

Objects Objects on construction projects are tools, equipment, and materials.

All three, along with the labor of the craftsman, must be present in order for production to take place. The tools and equipment are objects that remain permanently on the job site. They are available to the craftsman whenever he requires them. Materials, on the other hand, must come to the construction site in a continuous stream during the life of the project. The movement of materials to the project and to the work station is a social event involving teamsters, operating engineers, laborers, foremen, superintendents, clerks, and suppliers. Material movement and handling requires social communication between craftsman and foreman and then to the superintendent, field clerk, supplier, or manufacturer and trucker, in that order. The process then works in reverse as the material comes out to the job and into the field at the place of production. It becomes regularized and patterned as craftsmen call for materials needed to accomplish their work. It sets in motion a system that is repeatedly used to bring materials to the site when and where it is needed.

Concrete is an important construction object, composed of a mixture of sand, stone, and cement. Usually, a contractor will issue a purchase order to one supplier for all the concrete to be furnished on the job. The type of mix specified is submitted to the architect or engineer for approval. Sometimes, several different types are specified. When a concrete pour is made, the foreman tells the construction clerk how much and what type he needs and the clerk calls the concrete plant to order the material, giving the plant the time of delivery and location. There is a time limit within which concrete must be mixed and delivered, usually 90 minutes from the time it leaves the plant. If it is not poured within that time, it could be rejected by the inspector. Thus, concrete ordering and delivery must be planned, orderly, and systematic. Otherwise, losses and waste can result. With recurring concrete pour events, the ordering and delivery of concrete becomes routinized. Of course, as in all construction, things often happen to make plans go astray, such as a concrete truck break down, a heavy rain that interrupts a pour, or the plumber's forgetting to put his pipe sleeve in the wall.

Persons When a construction project begins, there is some awkwardness among the men while they get used to interacting with each other. Some are found to be unsuitable and are laid off. Others take time to learn the habits and personality types of the men in their crew. On the downtown road project, it took about five months for this period of adjustment to take place. The men in the various crews had to learn to work with their foreman. Shifts were made. An operating engineer did not get along with the labor foreman in his crew. The labor foreman on the bricklayer crew did not get along with the bricklayers, so the labor foreman was sent to another job. I fired the bridge superintendent who was too abrasive in personality and only slightly competent as a superintendent. Some operating engineers seemed better suited to one machine rather than another and reassignments among the operators took place. After the five-month period, the men seemed comfortable with each other. They came to know each other's moods and work habits. Some of the crews stayed together for seven months until the following year's winter layoff.

Conflicts Recurring conflicts are an inevitable and not necessarily unhealthy aspect of a construction project. If they interrupt the work they are serious. However, if there is absence of conflict because of absence of contact and communication between the trades, that too is serious. Conflicts often arise when one craft pushes another to get its work done so the complaining craft can do its job. This is a healthy conflict because it moves the job.

Conflicts inevitably arise between the owner and the contractor. Both are interested in a speedy project. Finishing on time is mutually beneficial. But each side has its own perspective. The owner views quality as primary. The contractor is interested in high efficiency and meeting costs. Low cost and quality often conflict and so do owners and contractors. There are also personal conflicts. Men belong to different crafts and there is craft rivalry. The men work under arduous and hazardous conditions. They must work in trying weather conditions. This can cause irritation and short tempers. Some days the work goes badly, making everyone grouchy or angry. Still, construction men must accommodate one another, not only as members of different crafts but as individuals. Personal conflict is a recurring activity, but so is horseplay and verbal joking. This serves as a means of blunting and controlling hostility. It is a form of joking behavior as described and analyzed by A. R. Radcliffe-Brown (1952), who saw joking behavior as a means of preventing open aggression and ensuring a stable system of social behavior. It also serves as a symbol of group solidarity. Insiders are permitted to insult each other, call one another crude nicknames and physically joust with each other. But let an outsider try the same thing and they will be assaulted physically.

On one of the sewer projects, we had a labor foreman who was known as a "duker" (a good fighter). He was also mild-mannered and had a wonderful wit. He enjoyed verbal dueling and laughed easily. One day he was fooling around with our prize clown, Tse Tse, when a concrete truck pulled up. The driver became very impatient and wanted to get his truck unloaded. He saw Tse Tse and the labor foreman fooling around and mistook the latter's laughing nature. He insulted the labor foreman and told him to get to work. Words exchanged. Then the driver jumped from his cab shouting, "You wanna fight?" He put up his hands in a fighting pose and within a few seconds was lying on the ground in the same pose, his nose broken and his face covered with blood.

ADMINISTRATIVE ROLES ON THE CONSTRUCTION PROJECT

Project Manager The role of the project manager is that of an interpreter. His function is to convey the policies of management to the men executing the work. He plans the overall direction of work through progress schedules, which he transmits to the working force through the superintendents and foremen. The project manager plans for material deliveries and equipment needed to progress the work. He monitors costs and carries out administrative duties with relation to the owner and architect. The project manager is the highest ranking officer in the field representing the contractor.

The project manager plays an interface role between the owner and contractor.

He carries out the policies of the contractor, and conveys to his superiors the requests and directives of the owner. The project manager overlays the work function with the administrative function. Project managers depend on the journeymen for the execution of the work and on top management for policy decisions. They are often compelled to act against their better judgment in carrying out their employer's policies. In this respect, they are less independent than journeymen, who are usually the sole judges on how to carry out their tasks.

In an article in *Highway and Heavy Construction* (March 1978), the editors stated that "it is the men who make construction companies go." Any project manager who ignores that will be ineffective. In industrial production where technology controls the movements of men, it makes little difference what a man's attitude is if he performs his limited set of operations each day. Attitude is crucial in construction since management does not control the operations by which the craftsman carries out his tasks. Thus, a project manager is only as good as his ability to create and maintain the kind of attitude among his workers and supervisors that will make them cooperative and productive.

Resident Engineer, Architect Representative, or Engineer-in-Charge The resident engineer, architect's representative, or engineer-in-charge represents the owner on a construction project. His main function is to ensure that the contractor will perform work in accordance with contract plans and specifications. In order to carry out this function, the engineer or architect employs inspectors who oversee the work and pass judgment on whether it meets the plans and specifications. The engineer or architect evaluates the contractor's requests for payment and approves, rejects, or modifies such requests as they relate to the amount of work performed by the contractor each month.

I have selected two models of engineers to illustrate the different styles of carrying out their duties.

Engineer No. 1 was a take-charge guy who often claimed he, and not the contractor, was "running the job." He wanted to play all roles, contracting as well as engineering. He ran his office as he saw fit, allowing no interference from his employers or the owners. He made decisions and stood by them, right or wrong. He was egotistical and boastful, believing he knew all there was to know about construction. In spite of his arrogance and boastfulness, the men respected him for his strength and independence.

Engineer No. 2 was a strong contrast to Engineer No. 1. Rather than taking a vigorous role in the work process, he preferred to stay in his office and not be bothered by problems. He left the work of solving engineering problems to his assistant. When he had to make a decision, he wavered for a long time and often rendered an ambiguous one. He took a strong stand on unimportant matters. He once threatened us with a fine because a sign was crooked. He was not respected even by his own inspectors and assistants. Though he was a good engineer and a knowledgeable man, he was unpredictable and unreliable as a personality. For weeks he would be completely silent. Then he would suddenly assert himself on a petty matter. He'd be friendly one day and sullen the next; helpful in the morning and obstreperous in the afternoon. He had none of Engineer No. 1's boastfulness. He was retiring and soft-spoken. He often remarked that he wished he had gone into another profession. He was envious of those with less education who made more money. He hated to write or receive letters because they became part of the "record" that some superior might read.

Both engineers wanted to have their projects built well and in accordance with the plans and specifications. For Engineer No. 1, he saw such a result as a personal triumph, something that would further his career. To Engineer No. 2, the main concern was to have a "quiet" project with no fuss or problems and very little paper work.

Superintendent The superintendent is probably the key person on any construction project. He is the link between the intent of the designer and its execution. He serves as interpreter in translating plans and specifications for the men he directs. The superintendent is the interface between management and the construction worker. He is also the liaison between crafts. Incompetency can be overcome or compensated for with almost any other construction role, but not with the superintendent. A poor or unqualified superintendent inevitably leads to a bad project.

The following is an outline of the style of the superintendent on the library project. This man is considered the finest superintendent in the company and thus, his supervision style may be considered a model:

1. His overall stance is democratic and consultative. Most times he tells the foreman what task has to be accomplished and permits the foreman and journeyman to organize the work in their own manner.
2. If something goes wrong he tries to find out why. He does not humiliate the man who made the mistake. He gets angry only if the man lies or tries to cover up an error.
3. He responds to a foreman's or worker's request as a "now" problem. If he has to say "no" to anyone, he says so clearly without anyone getting the wrong impression.
4. He never leaves the job site until every man who is working completes his tasks for the day.
5. He involves himself with paper work in a minimal way, permitting the construction clerk to handle most of it.
6. He tries to conduct all business face to face, preferring to talk to a man on the job rather than over the phone.
7. He drinks and socializes with the men who work for him.
8. He makes it a practice to have regular social events with the men from all crafts on the project.
9. His files on the job are disorderly but he manages to find what he needs.
10. He listens carefully and has a remarkable memory of what was said to him.
11. He is never intimidated by threats.
12. He does not hesitate to tell friends they are wrong if he believes it to be true.

Foreman All foremen in construction are working foremen. The foreman gets extra pay, but it is very little above scale, 30 to 50 cents an hour more. The foreman is close to the men he leads. Every crew needs one man to direct the others, plan the work, and lay out the material. This is the function of the foreman. The other men are too busy working to plan ahead. The foreman is in touch with the superintendent and thus finds out how the work of his own crew fits in with the rest of the project. The foreman plays an interface role between the superintendent and the workers. He carries out the overall strategy of the superintendent, assigning tasks to the members of his crew in accordance with the superintendent's goals.

Pete is a concrete finisher foreman. Like most construction foremen, he has a following of men he can assemble quickly if they are needed for a concrete pour. Pete talks about his men as if they work for him, not the contractor. There is a mutual loyalty between Pete and his men that overrides any sense of obligation the men have toward the contractor or even the union.

One of the norms in construction is that a foreman takes responsibility for all successes as well as mistakes made by members of his crew. An incident occurred on one of the jobs in which the foreman violated this rule and was caught. The foreman was a friend of the superintendent. Nevertheless, he was called "a fuckin' liar" and embarrassed himself when he tried to blame his men for work he had done which was sloppy and unacceptable. The point was that even if his men had done the work, the foreman was still responsible since he is supposed to oversee the work.

Foremen have a broker function. They make arrangements with various superintendents to perform work and then serve as employment brokers, getting employment for the men in their crew. Sometimes, crews under a certain foreman move from job to job as the need arises for their specialty.

Journeymen The journeyman's role in the main is his craft role. Management has its own ideas about what a craftsman's norms should be. They are adhered to by the journeyman if they do not conflict with his own craft norms.

The journeyman's style is to maintain his independence and keep control over evaluation of his work within his craft. He will accept the right of the superintendent or foreman to tell him where, when, and what tasks to do, but he will not allow anyone to tell him how to perform his craft.

Joe is a journeyman operating engineer. He describes himself as a "hardhead" because he insists on always doing things his way. Sometimes his employer tries to tell Joe how to do his work. Joe gets indignant, considers such interference presumptuous, and becomes almost hysterical in his rejection of such interference. When Joe talks about his work, he sounds more like a contractor than like an employee. He says that he has been with the company 15 years and that he quits every month and gets fired every other month. Joe enjoys getting his employer's goat. Once he broke a gas main and his boss asked him, "Why the hell did you do that for, Joe? You got our name in the paper, besides." Joe answered with a straight face, "Oh, it was a dull day and I figured you needed some excitement."

Construction work also includes the demolition of old buildings, which is a sad sight to see, particularly if the building had some architectural style and a graceful form.

A BUILDING DEMOLISHED
STATION 7 PLUS 14 R

Four stories high,
Wearing the morning's gray light
On its shoulders,
Majestic and solitary,
Standing as if waiting.

Boomed above, wheeling slowly,
The steel arm, its iron claw
Opens and clamps
A symmetrical gap in the brick
Opened now to the sky.

The building seems imperturbed
While its sides fall gracefully.

Jaws tug at the cornice,
Trimmed with Corinthian motif,
The heavy stone resists,
Three times the teeth bite
Before the granite gives way
And carved petals fall
From broken flowers,
Buried beyond recovery or remembrance.

The structure falls at its own feet,
Soon to be pushed into the earth,
To harden a place for a while
Before grains of stone and earth
Become one.

5/Independence and autonomy

INTRODUCTION

The content and significance of jobs involve two aspects. One is the technological and organizational requirements of the firm. The other is the social and personal requirements of the employees. Industry has concentrated on the first aspect. An extensive literature exists on the technological content of occupations. Technology is supposed to provide for rapid and error-free output of goods. The goal of industrial engineering is the adjustment of the work to the demands of technology. The capabilities of the worker are viewed in relation to work methods, machines, and equipment as they affect the physical aspects of production. The worker's role is seen as one element—although a human one—within the system of work technology.

Models of human behavior in productive organizations were passed on to the modern era by eighteenth-century economists (Heilbroner 1970). The most influential ideas that accompanied the rise of industrialism were the division of labor and man as an economic animal. The economic-man concept provided the rationale for our present reward system. The division of labor concept was the basis for the overspecialization of job and organizational structures. These approaches were suited to a period when the labor force was unschooled and unskilled.

During the late nineteenth and early twentieth centuries, business and industry employed a technology and social organization derived from the mechanistic model of human behavior. Taylorism was the practical expression (Taylor 1911). In Taylor's model, man's role is that of a cog in a complex production system dominated by costly equipment. In mechanical systems, all elements are completely designed. Initiative and self-organization are not acceptable, for they increase variability and the risk of failure. The result was rigidly specified task assignments and job descriptions indicating specific behaviors desired. The drive for reduced variability encouraged the idea of minimum skill requirements for task performance. Jobs were designed to be specialized and fractionated so they could be rapidly acquired with minimal training. To ensure successful outcomes, reward systems were designed to provide reinforcement only for precisely specified behaviors.

Based on the mechanistic model, modern business and industry retains for itself maximum flexibility and independence in how it uses its manpower. Business organizations strive for maximum interchangeability of personnel. The organization's goal is to reduce its dependence on the availability, skills, or motivation of individuals.

In mass production factory work, the product is not turned out by any one worker or group of workers but by the plant as a collectivity. Usually, the individual worker cannot define his own contribution to the productive process and say, "This is my work." The mass production principle is not confined to factories. Bookkeepers, typists, computer keypunch operators, supermarket clerks, and even professionals as members of large organizations are subject to the effects of mass production.

The above trends and practices have not taken hold in construction because the technology of the industry makes these practices inappropriate. The special features of construction work are the uniqueness of site and building; the temporary duration of the project; variability due to weather; and the inability to stockpile a bulky, fixed product. The organizational features of the industry are the skilled labor force; the small size of most construction firms; the control of the worker over the work process; and the strength of the craft union. In this chapter I hope to show that these features of the construction industry have resulted in independence and autonomy for the construction worker.

OWNERSHIP OF TOOLS

Most people living in preindustrial societies make their own tools and produce independently. This is true of hunters, farmers, herdsmen, fishermen, and craftsmen. It is not true in industrialized societies where there are few occupations in which people produce goods independently. Most individuals are dependent upon access to some organization or corporation in order to play a role in the productive process. In industrial society it is the business corporation, not the individual, which is the productive unit.

The divorce of the worker from his tools had enormous impact on the social structure and culture of industrialized nations. It was cited by Durkheim (1949: 233–255) and the papal encyclical, "Rerum Novarum" (Hayes 1946:306–307) as a major obstacle to satisfying work in capitalist nations. It was a main premise of Marx's explanation of worker alienation in modern society (Fromm 1961:44–58).

Construction is one of the few fields where significant numbers of workers still own their own tools. While this is not true of every trade, it is for several significant crafts—carpenters, masons, electricians, cement finishers, ironworkers, sheet metal workers, dry wall installers, plumbers, and others. Through ownership of tools, construction craftsmen have the choice of working for themselves rather than others.

Sometimes, ownership of tools and equipment by a journeyman makes an employer dependent upon him. On our road project, one of our bricklayers owns his own truck, tools, and some equipment. Though on the payroll, he acts like an independent entrepreneur. After he is given his assignment in the morning, he makes his own arrangements for material and equipment. He loads it on his truck, picks up his laborers, and drives to his work station. He owns his own scaffolding, heaters, and accessories, and if the contractor does not have these items on the job he "lends" them to his boss. A large number of men on the road project have their own trucks and use them to move tools and materials. Many of these men have the feeling of independence that comes from the "freedom to determine the techniques of work and to choose one's tools" (Blauner 1964:43). Thomas Brooks, labor historian, comments on the feeling that possession of tools gives to the construction craftsman (1979:8): "The construction worker's pride in his craft and his autonomy is . . . rooted in ownership of his tools, through which he symbolically owns his job and controls his destiny."

As previously mentioned, ownership of tools gives construction workers the alternative of going into business for themselves. Many have been in business. Many perform small jobs on the side on weekends or at night. Construction is an industry of small firms and working bosses. It requires little capital to go into business, often no more than some hand tools, a skill, and a contract. A mason who had been in business for himself said all you need to get into construction is to have a pen (meaning, get a contract). The chance of going into business is dwindling for most Americans. Today, 70 percent of Americans are employees of industrial and service companies; 15 percent are government workers; only 9 percent are self-employed; and 6 percent are farmers (HEW 1975:21): "As the data attest, the trend is toward large corporations and bureaucracies which typically organize work in such a way as to minimize the independence of workers."

The aspiration to be in one's own business is strong in the United States. A poll showed that 51 percent of Americans had this as their goal, compared to 33 percent of Britons (Cantril 1951:528). Blauner quotes an auto worker (1966:xvi–xvii):

> The main thing is to be independent and give your own orders and not have to take them from anybody else. That's the reason the fellows in the shop all want to start their own business.

AUTONOMY IN THE WORK PROCESS

Construction workers, particularly the skilled craftsmen, exercise autonomy over their work. Construction management relies on the tradesman to fabricate the product. Construction workers rely on themselves, rather than management, when confronting problems. The skilled craftsman professes to know his trade better than anyone else. He believes since he has the skill he should control the work. Management expects skilled men to do their work without the need for strict supervision.

Sometimes, management may have goals that conflict with those of the tradesman. It believes it has the "big picture" and insists that the craftsman take short cuts to speed up the work. The craftsman, however, lays stress on performance and considers speed the enemy of quality.

Quite often, if a construction man considers the working conditions onerous, he will not hesitate to quit. He has enough self-confidence that he can find another job. At a company dinner, one of the men quipped that he should organize a "second-timers" club. He saw so many men around him who had quit the company and were later hired back.

Mike Cherry expressed the attitude in this way (1974:183):

> A casual dislike of the boss is probably not a central factor in a white-collar's decision to leave his job, but it is often the sole issue in whether or not a construction worker stays on. If the boss wants a thing done one way which the worker thinks should be done another, or if the worker thinks the boss has spoken to him in an unacceptable tone of voice, or if the man simply disapproves of the way the captain runs the ship, he is likely to make a "shove it up your ass" speech and leave.

CONSTRUCTION TRADE UNIONS AND WORKER AUTONOMY

A significant portion of union agreements in the construction field deal with the prevention of encroachment by management upon worker autonomy. Building trades unions utilize the closed shop to restrict management control over workers. The carpenter's union agreement states that a "commercial qualifying card" shall be a condition for employment. This card can only be issued by the carpenters district council in a particular area.

A major tool of management encroachment on worker autonomy, supervisory personnel, is controlled by the union. Foremen and superintendents are union men. As such, they usually feel more loyalty toward the men and their union than toward management. If a foreman violates a union regulation or becomes autocratic, he can be fined or brought up on charges before the union executive board. All this severely restricts management's power to control and dictate to the men.

Construction unions have other means to protect worker autonomy. The unions, not management, control job entry into the industry since, in most cases, a man must be a union member to get a job. The unions also control the apprenticeship program. They administer the schools and provide candidates for apprenticeship. Employers sit on apprenticeship boards, but it is the union that has the final say on whether or not a candidate will be accepted into the union and the craft.

Union agreements also deal with union encroachments on management autonomy. Unions are not supposed to place any restrictions against the use of machinery, tools, or labor-saving devices. They are also not supposed to place any restriction on labor output. In practice, there are restrictions on how machinery is to be used. There are also informal norms by each trade on the amount of work a journeyman should produce. Trade unions, often blamed for restricting technological improvements, reflect the attitudes of their membership. In resisting technological change, they are trying to prevent the reduction of worker control and autonomy over the work process.

CONTROL OVER THE NORMS OF PRODUCTION

In a work on construction management, O'Brien and Zilly (1971:10) state that regarding labor output in construction, "on the local level there are still unwritten laws that govern the output." The approach to estimating costs provides an illustration of construction worker control over output norms. Our chief estimator told me he keeps "historical cost data" and estimates labor costs "after the fact." When the company bids a project, cost figures are examined to ascertain what the men are currently producing. This is used as the production norm. The chief estimator said he might try to calculate improved methods for expediting materials or might consider using a less expensive material, but labor output was beyond anyone's control other than the journeyman.

Worker control over production norms is not as much a question of "restricting output" as "deciding it." At the sewage treatment plant, there were several occasions when the men placed a thousand yards of concrete in a single day. No one from management told them to do it. The superintendent, foremen, and key journeymen decided and planned it on their own initiative. One evening, at the local bar, in July 1976, after one of the thousand-yard pours, Pete expressed the pride and satisfaction that comes from extraordinary accomplishment, and said:

> If they'd leave us alone, we can take care of the work and make money for the company. We did a thousand yards today. But I've done better. As long as Carmen [his employer] leaves Earl [the superintendent] alone we can turn out the work.

Another feature in construction which strengthens worker control over production is the personal familiarity between management and workers. Field supervisory and management personnel come from the ranks. Often, these men have worked with or been friends with the men they supervise or employ. This familiarity permits the men to assert their independence over how and how much they can work. A pipe foreman once remarked about his boss:

> He and I used to lay pipe together, just like this, in the trenches. Now, whenever he sees me working, he tries to tell me how to do it and how many feet I should get. I just tell him to get lost and do my work.

RELATIVE DEMOCRACY

On construction projects, within each trade and among the trades, there is a sense of democracy. No particular trade is considered superior to any other. Every trade makes a contribution without which no building would be complete. The painter of a wall is no less important than the carpenter who built it. The window caulker is just as vital as the glazier. No mason could lay block if his laborer did not mix the mortar and erect the scaffold. Within each craft each journeyman gets the same wage. The wage differential between trades is very little. One trade may command slightly higher wages, while fringe benefits and conditions are better for another. Men are respected because they perform quality work, not because their trade ranks higher.

Construction men consider themselves the equals of others, regardless of rank. Engineers and employers are treated with respect, not deference. Every man on the job is expected to carry his load. There are no prima donnas because of rank. At one of the sewage treatment plants, an older laborer was given the clean-up job because of his age. He also ran errands, picked up small tools, and performed odd jobs. No one looked upon his work as menial nor looked down upon him. He was so respected that he was made labor steward.

REFUSAL TO WORK IN BAD WEATHER AND DANGEROUS CONDITIONS

In most industries, refusal to work can be cause for dismissal. However, in construction, the workers, not management, determine which conditions are unsafe or unsuitable to work in. The carpenters' agreement, for example, states that "employees will not be required to work in inclement weather." The carpenters' shop steward and the men themselves jointly decide if the weather is inclement. This type of decision making by workers in factories does not exist. It prevails in a few sectors of industry—among miners, longshoremen, truckers, sandhogs, fishermen, and other high-risk trades. Construction men are paid "show-up" time if they report to work and the weather becomes unsuitable.

There is the possibility that a superintendent may not agree with the men that the weather is too bad to work in. Most superintendents are union men and wish to maintain their good standing with the men. There was an instance on one of the sewer plant projects when a superintendent opposed the men on whether or not to work. The superintendent refused to pay "show-up" time. He claimed the men could have worked. The company had to back him up, but behind the scenes they criticized him. The superintendent continued his arrogant ways and alienated the men further. He was finally taken off the job and sent to another project. He had trouble there, too, and eventually left the company.

Construction workers will not work under unsafe conditions. They will not go into a trench that is unshored. A crane operator will not lift a load if he believes it is too heavy for his machine. The government protects the right of workers to refuse work under unsafe conditions. A man fired for refusing to work when it is unsafe may appeal to the federal Office of Safety and Health Administration (OSHA). A project manager who orders men to work in dangerous conditions can be fined and jailed if it results in injury or death (*Engineering News-Record* 1976:16).

Jack Haas (1977:166) describes a situation with high steel ironworkers who determined for themselves that it was unsafe to work and refused to do so:

> The decision to work or not may be a group one; when a decision is made to thwart the foreman or contractor's usurpation of their autonomy the group will

invariably act together. In one example, the foreman directs the workers to shovel snow off an iron beam; the group chooses not to work, and the men leave the job. Other workers remain free to decide for themselves whether or not to work. The perspective of worker autonomy is enacted and reinforced.

AUTONOMY IN HIRING AND FIRING

Hiring, firing, and layoffs usually take place on the construction job site and not in the company office. Men seeking work contact the superintendent, foreman, or shop steward on the project, and the home office never becomes involved. Otherwise, the superintendent or foreman contacts the business agent to have the union send men to the job. Hiring is also done through personal contacts. A superintendent or foreman needing men contacts those he knows personally at their homes and tells them to report for work. The officers of the contracting firm may express concern about the numbers of men on the project, but rarely do they participate directly in the hiring and firing process. Commenting on the autonomy of construction workers as a result of the hiring process, Richard R. Myers writes (1946:1):

> Perhaps more than workers in any other industry the building workers function as independent units, each worker pursuing employment and making arrangements to apply his skills according to personal contacts, personal preference and a personal schedule.

Bennie Graves studied construction workers in the pipeline industry and found that autonomy of the workers was similarly related to their autonomy in hiring and firing. Men were hired through contacts with relatives or friends or made their own "deals" with superintendents and foremen. The superintendent became dependent upon the men in the summer, when work loads were high and manpower scarce. The men were dependent upon the superintendent in the winter when work loads were low and manpower high. This reciprocal dependency placed the workers on a relatively equal footing with the men who hired them. Graves describes the relationship (1974:422):

> Autonomy requires that networks (workers, foremen, superintendents) rather than "outsiders" (unions, employers, public officials, pipeline owners) must control most day-to-day aspects of hiring, firing and conditions of work.

A man who secures his job through his own efforts is independent even of the union. A labor foreman talked about this at a company's annual Christmas party. He said that since he secured his job through his own contacts, he could hire his own men and form his own crews without going to the union. He contrasted his situation with that of another foreman, who secured his job through the union and was obligated to take any man the union sent him. The first foreman said, "I got my job myself so I owe nothing to the union. I get my own crews and so they owe me. The man who hired you or helps you get your job is special."

It can be generalized that the hiring and firing process in an industry is an important clue to the level of autonomy of the workers in that industry.

AUTONOMY AMONG GLASSBLOWERS

Construction workers are not alone in enjoying autonomy in their work, and it is instructive to compare them with other occupations that have work autonomy. In a study of glassblowers and shapers, William F. Whyte (1974:Chap. 10) describes an occupational group in which the "social work process allowed much freedom" (1974:149). Whyte bases his study on the technological aspects of glassblowing. He points out that while there has been a great deal of chemical engineering applied to the development of the glass material, the fabrication of the glass itself is done with methods that are centuries old. Therefore, the glassblower of today enjoys an autonomy that stems from the work process, although the degree of freedom has been modified by modern conditions.

Whyte's study was done in a factory that employed more than 6000 workers. One portion of the factory was devoted to producing glass objects of artistic and esthetic quality, some of which were engraved. The working team of men included a "gatherer," who picks up a piece of molten glass on the end of a hollow blow iron and rotates it; a "servitor," who shapes the piece close to its final form by blowing and using hand tools; a "stick-up boy," who carries the piece at the end of a rod to the "gaffer," who is the most skilled of the team and shears, shapes, hand tools, and blows it into its final form.

In the old days, the gaffer was the autocrat of the team. He selected all the members, and anyone he did not approve of did not work. He directed the work tasks of each member, and whatever training he chose to offer to the servitor (who was next in line to becoming a gaffer) was strictly at the gaffer's discretion. The old-time gaffers were almost all Swedes. They said that it took 20 years of training for a man to become a skilled craftsman. The company permitted the gaffers to have complete freedom over the esthetic ware shop; in some cases, if a gaffer did not feel like working, or was having a bad day and not able to form his pieces just right, he might just go home, which left the other members of the team no other choice but to do likewise. Advancement to the other skilled positions was slow and was under the control of the gaffer.

All of this changed with modernization and unionization. The plant set up a visitor's gallery and found a much larger market for its artistic ware. It went to two shifts and found it needed a larger number of skilled craftsmen. The entire plant was reorganized in 1941, and the selection of training and the mobility of job process passed out of the hands of the gaffers and into the hands of the union. The old-time Swedes began to die off and were replaced by Italian-Americans and other ethnic groups. The union took over the control of training skilled craftsmen. It was no longer possible for a gaffer to reject a man he did not like if the man was qualified for his craft. In spite of these social changes, the skills remained basically the same as they had always been, except that more and younger men were now being trained for the various crafts.

Even with the changes that occurred with modernization, the gaffers continued to be the controlling force on the work team. When problems arose, it was the gaffer who took the lead to correct them, not the engineering department of the company or some person in management (Whyte 1974:155):

> Jack Carter (a young gaffer) met technical problems through directing a group discussion among the members of his shop. Whenever any new problem came up, he would call the group together for a pooling of ideas.

A number of conflicts arose when attempts were made by management to encroach upon worker autonomy. Efforts were made to apply time and motion studies, but they were quickly abandoned. The company management established production norms, but the men invariably produced less and complained that the requirements were too high. At one point, the plant manager thought that by placing a foreman on the floor where the men were working, the foreman could help with technical problems and productivity might go up. However, since the foreman knew nothing about the skills involved and was merely an administrator, his presence added nothing. Besides, the men knew why he was placed on the floor and resented it. Management, realizing it was futile to keep him there, withdrew him from the floor. Management learned most pointedly that they had little control over the work process when the men "fired over." If a crew was having a particularly bad day and the product was not turning out as it should, the entire crew would leave the job for the rest of the day and begin anew the following day. The company neither fired the men for walking off the job nor did they dock them in pay.

Whyte's conclusion about glassblowers was that they enjoyed freedom from control by management over the execution of their work. Their work relations were influenced by new technology and management action to a limited degree. Unionization and a changing social environment had a greater effect on the relationships between the men. The work situation permitted these men to enjoy a high degree of responsibility and creativity in solving their work problems. Job satisfaction was very high among them.

Comparing construction workers with glassblowers, one finds similarity in their control over the work process. Also, management blindness to field or work problems is as prevalent in both occupations. Just as Jack Carter called his crew together, we recently (July 1978) had a meeting in the field on the road project to try to figure out why our compaction tests were failing on our earth embankment work. After discussion and investigation, we concluded that our vibratory roller that compacted the earth was too small. We conveyed this information to top management, which refused to accept our assessment of the situation and told us to continue to use the same equipment. Meanwhile, the state inspectors were threatening to shut down that particular operation if we did not get larger equipment on the job. After a week of continued failing tests, the operation was closed down, and top management was forced to purchase a larger piece of equipment. They bought a used vibratory roller after delaying for another week. The roller broke down as soon as it came to the job, which caused a third week's delay. Management's refusal to act quickly based on the assessment from its field personnel caused a three-week delay, which in turn caused a delay in building the bridges associated with the earth embankment work.

REDUCTION OF AUTONOMY AMONG RAILROAD WORKERS

Railroad workers in the United States at one time enjoyed high prestige and status among the working classes. They were among the aristocrats of labor, enjoying high wages, autonomy in their work, and an independence of spirit that set them apart from other workers. In fact, when they were organized, the Railroad Brotherhoods maintained their independent status and did not affiliate themselves for a long time with either of the major labor federations. Through technological changes in the industry, major changes occurred in the social organization of the industry (and the labor within it) to reduce the large degree of status and independence formerly enjoyed by railroad workers. In this sense, the railroad workers are dissimilar to construction workers, who have not had to suffer major technological changes that reduced employment and affected the social organization of the industry. Furthermore, employment in the railroad industry is declining, while construction employment is rising. But it should be instructive to see what effects technology has had on railroaders, compared with the situation among construction men (Cottrell 1951).

One of the major changes that reduced worker independence was the change in engines from steam power to diesel. The old steam locomotive engineer was a legendary person, about whom stories and songs were written. He was said to be the master of the engine. It took many years to train a good engineer and years of experience to learn the turns and twists of the "road." A good steam locomotive engineer was said to "feel" its power and controls and to make the necessary adjustments based on his experience and knowledge. The new diesel engines are more automatic. They have automatic control systems that eliminate the need for the engineer to "feel" the controls. The efficient operation of the engine depends less on the ability of the engineer and more on automatic control systems. Diesel lubrication must be automatic and not dependent upon the human "oilers" used for the steam engines. Even the skills required for repairs changed with the switch from steam to diesel. The mechanic on the steam engine needed skills for fitting the parts of the heavy equipment, whereas the mechanic on diesel engines removes parts and replaces them. However, the diesel mechanic must have wider knowledge than the old steam mechanic in the area of electrical work and electronics, which are involved in the control systems on the diesel locomotives.

Another aspect of railroad technology which changed the organization of work was the introduction of the long welded rail in lengths of 1000 feet in place of the traditional 30- to 40-foot lengths. In the old days, men could pick up the rail with bare hands or a pair of tongs. They would spike the rails to wooden cross ties. Now, the long rails are handled by machines which also fasten them to cross ties and tamp the ballast beneath. The machine has largely replaced the human hand in rail laying and in rail replacement and repair. The signaling technology in railroading was another instance of the replacement of human labor with mechanical devices. Instead of the Morse code operator and the flagman, railroads now use computers and automatic signaling and switching of trains from a remote control center.

At one time in the United States, the Pennsylvania Railroad Company was the most powerful corporation in the United States. It reflected the position and power of the railroads in the nation. During this period (1900–1920), railroad workers enjoyed a status and independence second to no other group of workers in the country. They had their own independent union organizations, they were among the highest paid workers, and they had their skills and crafts on which to base their prestige, power, and economic benefits. With the advent of technological changes in the industry, the railroad workers lost jobs as well as prestige. Between 1920 and 1970, the number of railroad workers fell from 2 million to just over .5 million. It is very difficult for an occupational group that loses such an enormous percentage of its members to technological change to maintain high prestige and standing. It takes on the character of a dying occupation, one fraught with problems of insecurity and filled with obsolescence. Thus, railroad workers, as compared with construction workers, are in a contracting labor pool, whereas construction workers are growing in numbers. Furthermore, railroad workers are losing their autonomous status and skills as a result of technological changes, whereas construction workers have not been subject to drastic technological change and have maintained their independence and autonomy.

AUTONOMY AND CONSTRUCTION WORKERS IN OTHER COUNTRIES

It is instructive to compare the degree of autonomy among construction workers in the United States with other countries. I have chosen two contrasting societies to examine and compare with American construction workers. In China, the construction industry is under the direction of the central government. British construction work is performed under a free enterprise system.

The material on Communist China is taken from a work by Kang Chao (1968). Since 1949, construction in China has been under the direction of the central government. At first, authority was placed in the various ministries that controlled the economy. This led to much overlap and bureaucratic mismanagement. For example, one ministry might build a plant in a particular area by moving the construction workers under their jurisdiction into and out of the area. Another ministry building in the same locality would then move a whole separate group of construction workers into the locality and then move them out when the project was completed. This costly overlapping led the Chinese government in 1960 to a decision to base the construction industry on local areas rather than ministries. Thus, it is interesting that Communist China and capitalist America both organize their construction on local area rather than nationwide industry.

Another similarity between the Chinese construction industry and that of the

United States is the use of the specialized subcontractor firm. The Chinese found that when they tried to let the general contractor employ all the labor, the general contracting firm did not have sufficient expertise to direct the men. The use of the specialized subcontractor, as well as the decentralization resulting from the organization of the industry according to the local area, mitigates somewhat the strong central control by the government and allows some degree of autonomy by local areas and specialized firms. However, the size of the Chinese construction firm, as compared with the United States, is enormous. The average Chinese firm employs from 2000 to 8000 workers, while the average construction firm in the United States employs less than 50. Being part of such huge organizational structures would hardly be conducive to worker autonomy or personalized relationships that characterize the typically small construction firm in the United States.

The number of skilled men in the Communist Chinese construction labor force is very low. This is a result of the newness of the industry and its great expansion since the revolution. The median age is very low (under 25) as compared with 45 in the United States. With a low percentage of skilled craft workers (30 percent, contrasting with over 50 percent in the United States), the kind of autonomy that comes with craftsman control of the work process would not be present in China's construction industry. The shortage of skilled labor and subsequent lack of autonomy leads to the use of draft labor, which is mainly unskilled. Labor can thus be drafted and assigned to construction projects in various parts of the country.

Mechanization and standardization are both strong trends in the construction industry in Communist China. In the United States where the firms are small, most lack the capital for investment in large construction machinery. Moreover, the machines are specialized; whereas a man can dig a ditch and then be put to work unloading a truck, a ditch digger can only perform one operation. Thus, utilization time becomes an important consideration in investment in large construction machinery. Since China has a large supply of cheap labor, mechanization can often be more costly than using human labor. Mechanization is important where there is a large supply of skilled labor; in order to reduce costs, machines which can be run by unskilled labor or which replace skilled labor are used. This is not a factor in China. Furthermore, China is experiencing a shortage of competent operators to run their construction machines.

A more important trend in China is the use of standardization of building materials and designs. In a free enterprise economy, where demand is unstable and fluctuating, as in the United States, standardization is very difficult to introduce. But in a governmentally controlled economy, where what is produced is determined by the economic plan and not by consumers buying in an open marketplace, standardization can be decreed and put into practice by the government; the consumer has to take what the government has decided to produce. Thus, standardized parts and building designs were introduced in China, as it has previously been done on a large scale in eastern Europe, Korea, and the Soviet Union. With standardization and the shift from the use of field fabrications to the use of factory produced units, there is a reduction in the use of skilled, craft labor and in craftsman autonomy. Building components come to the job and are assembled with semi-skilled and unskilled labor according to a preplanned specification and design that leaves little room for worker decision making or problem solving. Organization becomes routine and regularized, and factory-like conditions are imposed on the construction process.

In conclusion, there is little autonomy for the construction worker in China. The centralized nature of the industry, the large size of the construction firm, the use of draft labor, and the emphasis on mechanization and standardization are all trends that reduce worker autonomy. There is also the general atmosphere of a centrally controlled economy and an autocratic society run by the Communist party, which is not the kind of environment in which worker autonomy flourishes.

The construction industry in England is very similar to that of the United States (Colclough 1965). The significance of construction in England is greater than in the United States—one in ten people are employed in British construction. Including the building supply industries, construction accounts for one-eighth of England's gross national product. Colclough points to the jockeying of position in the British construction industry, with some firms switching from general contractor to one of the specialty subtrades and vice versa.

> The industry is one which depends on the skill, competence, enterprise and ambition of the individual rather than any possession of capital assets and fixed equipment (Colclough 1965:6).

Since the individual craftsman is of great importance in the industry, autonomy of the construction worker in England is high. Because it is an industry of small firms and low capital investment, most businesses in construction are dependent upon the energy, skill, and personality of the individual rather than the organizing skill of a large bureaucratic organization. In England, 25 percent of all contractors have no employees; thus, they depend on the organizing ability, skills, and autonomy of the individual working boss. Of those firms that do hire employees, seven out of eight have less than ten men working for them. The small firm, where personalized relationships and worker autonomy prevails, is more pronounced in England than in the United States. Colclough notes the high percentage of proprietorships rather than companies in the construction industry of Britain, which attests to the fact that the small contractor is willing to take personal responsibility for the quality of his workmanship and contracting management. Colclough points out that the small proprietor has a tendency to keep a permanent work force and to plan his work so as to maintain a constant level of employment irrespective of weather conditions (Colclough 1965:24). Men who work steadily for a firm tend to perform higher quality work, since they will still be around when the nature of the work comes to light at a later date. Small firms have little use for the unskilled worker and prefer skilled craftsmen on whom they can rely for varied operations and high-quality work. These craftsmen are accorded a large degree of freedom and autonomy in the work process, and since they are given steady employment, they reciprocate with commitment and quality work (Foster 1969).

As in the United States, in England it is the skilled craftsman who ultimately controls productivity through his control over the work process. Colclough states (1965:76) that productivity techniques in construction have evolved over thousands of years during which craftsmen have discovered for themselves the most efficient and economical way of carrying out every movement, and in fact, since Roman times there has been little left to discover. Except for the language difficulty an ancient Roman mason or bricklayer would be quite at home on many present-day building sites.

Another strong similarity between the construction industry in the United States and Britain is the strong personal relationship between employers and their employee-craftsmen, a characteristic associated with autonomy for the worker. Colclough remarks (1965:128):

> Employers and operatives in building work have a closer understanding of one another's problems and capabilities than is to be found in any other industry. Every craftsman sees himself as a potential employer and in fact a large proportion of those entering the industry make an attempt at some stage in their industrial careers to start their own business.

COMPARISON OF AUTONOMY AMONG WHITE-COLLAR AND CONSTRUCTION WORKERS

Traditionally, white-collar workers have been considered to have a large measure of autonomy in their work situation. They work in offices where they are close to management (if not their employers themselves), and they work in a relaxed atmosphere where there is no assembly line or close supervision pushing them to produce physical products. Furthermore, the work is not mass production oriented, although there is much routine in the nature of the work tasks. With the growth of large business organization and large office staffs, white-collar workers find themselves working in an atmosphere similar to factory workers in that they are surrounded by hundreds of other workers performing basically the same tasks. This is true of the lower level of office workers—clerks, file clerks, typists, computer operators, bookkeepers, keypunch operators, and others. As an example, we shall examine white-collar workers in six large insurance companies.

Michel Crozier carried out a study of white-collar workers among six insurance companies in Paris, France (Crozier 1971). Unlike construction workers, among whom there is a high degree of camaraderie and cooperation, Crozier (1971:110) found that office workers

> prefer a certain distance be maintained. Eighty-five percent of them never get together with their colleagues outside of work, and the 15 percent who do seem to apologize for it. The general order of the day seems to be "every man for himself," "we see each other enough during the week."

Crozier points out that white-collar office workers are attracted by the culture of their employers. Those he observed worked in a large city—Paris. There the commercialized culture was available to them. They felt they were part of bourgeois society, "which they cannot oppose but from which they feel, on the other hand, constantly excluded" (1971:188). Construction workers, in comparison, consider themselves blue-collar workers. They have no illusions about being part of the culture of their employers. Construction workers explicitly differentiate themselves from employers and professionals. They believe that most professionals, employers, white-collar workers, and government workers who work in an office are pencil pushers who work very little and are not capable of putting in a hard day's work doing physical labor.

Crozier analyzes the several positions from which he drew his sample of 358 subjects (1971:70). He presents an analysis of their work as it relates to the question of employee autonomy. Crozier, like many other authors (see Mills 1976) believes that "interest in work and evaluation of position are similarly, and positively, related to the feeling of autonomy at work, to the impression of not being tightly controlled" (Crozier 1971:99).

Crozier found that file clerks were at the bottom of the social ladder among office workers. They not only had low wages and skills, but "are the most dependent of all employees. That explains why among them one finds the strongest demands for autonomy" (1971.81). To the question, "Do you believe that employees can by themselves develop the best way of accomplishing their work?" filers responded "yes" twice as often as clerks and ten times as often as policy men (Crozier 1971: 81).

According to Crozier, keypunchers do not suffer the same degree of personal subordination that file clerks do. File clerks are on call and must be available at all times. Keypunchers have a regular job to perform and "in relation to file clerks, they have about the same position as semiskilled machine operators in relation to laborers" (Crozier 1971:81). The constraint imposed on the keypuncher is that of productivity. In general, keypunchers work in small groups, about 20 persons in all. Their small work groups, their special skill, and the fact that they are often trained on the outside by the manufacturers of statistical machines give them a place apart in the insurance company and a certain degree of autonomy. The keypuncher, like the carpenter or mason, has an occupation that is extremely well defined, and is the same regardless of profession or company. Like the skilled workers in construction, the demand for keypunchers remains high in business, both in France and in the United States. For all of these reasons, keypunchers are much less subject than their coworkers to the pressure of the work norms of the company they work for, and they enjoy a fair degree of autonomy (Crozier 1971:82):

> They are at one and the same time more demanding and more detached—in a word, freer. Finally, the relationship they maintain with their company is essentially a financial one, with productivity exactly measured on the one side, and premium pay on the other.

Typists, like keypunch operators, have manual and mechanical experience and training. Unlike keypunch operators, however, they depend much more on the company they work for, since most are specialized in their experience and the kind of material they have to type. Their jobs require a certain type of experience which consists mostly of becoming familiar with certain models of letters, forms, procedures, and sometimes technical language that is particular to the company that employs them. Thus, they have a little more autonomy than file clerks but much less than keypunch operators.

Clerks form the largest group, both in banks and in insurance companies. They also constitute the most heterogeneous group. The basis for this group is not a common function as among file clerks, or a common technique, as among keypunchers or typists. More than anything, being a clerk signifies being defined as neither this nor that, but rather as belonging to a mass of multifaceted personnel who have developed a number of bureaucratic skills which are called for according to the demands of the work. Clerks are the ones who come closest to the traditional idea of the office worker, the white-collar civil servant, and the large mass of white-collar workers. In insurance they are the solid base of the system, and they develop the greatest seniority within the company. The policy writers and claims adjusters in the insurance companies change frequently, but the clerks seem to remain. Crozier comments on the diversity of the functions of the clerks, involving counting, collating, noting, checking, inscribing, recopying, and filling out forms. The following were the activities listed among 40 clerks interviewed (Crozier 1971:84):

> . . . verification work (receipts, riders, policies, memoranda, agent accounts, and the like); assorted filing; calculation of rates, premiums and commissions; establishing or analyzing various facts and clients' situations; keeping books; recording receipts and claims; keeping accounts; reproduction services; card-indexing; graphing statistical tables; minor correspondence; complaints; special contract work; cashier work.

The competent man among the class of clerks is the one who has been through a great deal of different types of work and experience. The same is true of the construction man—the most competent are those who have been on many different types of jobs. The most competent clerks are the most autonomous and the ones who are the most consulted for advice in solving problems. They are independent in that their knowledge is initiatory. With the introduction of computers and automatic recording and storing of information, the situation of clerks will change drastically, but at present the replacement of the clerk by the computer is still in its initial stage.

Policy writers and claims adjusters are at the very top of the reward and prestige hierarchy in the insurance field. There are a series of job graduations between specialized clerks, who have some law background, and qualified policy writers and claims adjusters capable of developing the procedure to be followed in important insurance cases. Policy writers determine which documents are necessary at the request of agents in the field, after taking into account rules and tables supplied to them. Claims adjusters handle the more difficult problems in the settlement of accidents and damages insured by the company. The work involves discussions, negotiations, and decisions, handled by letter or orally. The most competent claims adjusters, those charged with handling contested claims, often have law degrees.

Compared to other insurance employees, policy writers and claims adjusters have a privileged position with regard to independence and autonomy. Most of them have the feeling that they are doing work which is important and enriching (Crozier 1971:86):

> Moreover, these staff members enjoy a rather high degree of autonomy. They are "individuals." Each one handles his own affairs and must assume his own responsibilities. They are unanimous in declaring that they must have a professional outlook.

In his summary of the position and general outlook of office workers, Crozier states that they are in a middle position in that they are workers, yet they are attracted to the culture of their employers. Unlike construction workers, who will band together in the face of hostility or authoritarianism from their employers, white-collar and office employees do not resist encroachment by management openly but rather by indifference and apathy. The office worker operates in his role with attitudinal ambiguity. He has the semblance of independence and autonomy and strives to move up the hierarchy of positions in order to reach the upper level where autonomy and independence are permitted. Yet, he is aware that his chances of succeeding are extremely remote; this knowledge, in turn, serves to make him accept and even demand forceful and authoritarian supervisors. The construction worker, on the other hand, has no illusions of moving up the hierarchical ladder. Once he has attained the position of journeyman, he knows he can go no further, either in position or salary, except for what his union will negotiate for him. He could become a foreman or supervisor, but that usually rests upon his own choice, and most journeymen are content to remain where they are rather than assume the added burdens and responsibilities of foreman or superintendent. The construction journeyman, therefore, believes in his own autonomy and independence and will not accept any encroachments upon those privileges. Not only does he not ask for authoritarian leadership, he has the ability to reject it.

WORKERS IN MODERN FISHING: ANOTHER CASE OF AUTONOMY

In a study of work organization in modern ocean fishing, Norr and Norr (1978) found a series of technical and environmental constraints which resulted in consistent patterns of work organization that led to worker autonomy. The authors found "flexibility in fishing arrangements; fewer levels of authority in fishing; more limited scope of authority in fishing with more decisions made cooperatively and fewer decisions the exclusive prerogative of leaders" (Norr and Norr 1978:164). The Norrs also found that fishing had limited levels of hierarchical control (usually not more than three) and that the industry exhibited a nonbureaucratic structure of work organization. In general, the authors discovered that "workers maintained control of productive factors even in the face of increased levels of capitalization" (1978:164). Stressing the autonomy of workers in the fishing industry, Norr and Norr summarized (1978:165):

> We identified five technical and environment constraints of fishing that increase the relative power of workers: exposure to physical risks, uncertainty, separation of work-place from residence, difficulty of maintaining clear-cut control of productive factors, and the need for teamwork, skill and reciprocal coordination.

Each of these constraints increases the importance of fishing laborers for effective production.

It is striking how many of these factors are similar to the technological and organizational constraints in the construction industry: the exposure to physical risks and danger; the uncertainty element in the industry; the need for teamwork and reciprocal coordination between skilled craftsmen. The conclusion drawn by Norr and Norr about the importance of workers for effective production can therefore be applied to construction work as well.

EARTH LOADER ON ROUTE 33

Saffron trees bleed through the morning haze
The clouds—silver, gray, smoke,
Freeze grayness into the windless day.

The blunt earth loader
Grunts and bounces,
Gouging into the land,
Layering black soil over brown clay.

In the cab, the operator,
Absorbed in his task,
Unaware of the birds
As they dart and peck for worms
Displaced by the tireless machine.

Birds, indifferent to the man
As he is to them.

6/Accidents, danger, and death

The construction industry has the highest incidence of accidents and deaths than any other industry in the United States. The average rate for injuries and illnesses per 100 full-time workers in American industry as a whole is 8.5. The rate per construction workers is 15.7 (*Engineering News-Record*, Nov. 15, 1979). In 1978, there were 628,000 injuries and illnesses due to construction work in the United States. According to the Bureau of Labor Statistics, the construction industry accounted for 20 percent of all deaths in industry, but employs only 5 percent of the total number of employees (*Engineering News-Record*, Nov. 15, 1979).

Of the fatalities in the construction industry, 29 percent were related to falls; 16 percent to over-the-road motor vehicle accidents; and 13 percent to industrial vehicles and equipment. An additional 10 percent of the deaths were due to contact with electrical current, followed by 6 percent from falling or flying objects and 5 percent from fires and explosions. Heart attacks accounted for 8 percent (*Engineering News-Record*, Nov. 15, 1979).

Since the number and type of accidents are many and widely varied, I have had to confine the discussion to only a small percentage of the incidents that occurred on the four projects researched.

FALLS

A Slippery Fall One of the causes of accidents in construction is bad weather. When a job site becomes wet and muddy from rain or snow, it is dangerous. Men lose their footing, equipment slides out of control, visibility is poor, and workers are absorbed with their own states of discomfort rather than external events. Here is an example of what can happen.

David P. was working as a surveyor at a sewage treatment plant project. One day he was on a 20-foot wall with a concrete pouring crew. His job was to give the men top-of-wall elevations so they would know where to stop the concrete. It had been raining for nearly a week prior to the pour. Ladders, planks, walls, and the ground were muddy and slippery. David was on the wall when the lunch-time horn sounded. He covered his surveyor's instrument and started toward the ladder.

Reaching over with one foot to catch the top rung, his other foot slipped and down he went. He later recalled his thoughts during the fall:

> I started to go down and thought, "This can't be happening to me." I knew I'd be hitting concrete. I hoped I wouldn't break my spine or neck. Then I hit. Something cracked the back of my head. It was like a window shade shot down and everything went black.

David landed on his buttocks. He was lucky. But his head snapped back and struck the concrete. He remembered coming to:

> I saw my dad bending over me. I had no sensation. Not anywhere. I figured I hit my spine. I couldn't even move my fingers. I was panicky. "My God," I thought, "I'm paralyzed." I was scared shit.

So was everyone else. No one knew then what the damage was. Someone called an ambulance, which soon arrived and roared off with David to the hospital.

Figure 7. Welding reinforcing steel on the inside of a form for a reactor vessel. After the reinforcing steel is in place; another form will be placed on the outside and the concrete poured between the two forms. The reinforcing steel is lapped over each other and welded together. The steel, inside the concrete in tension, strengthens the concrete. It can be seen from this picture why falls make up the largest number of accidents and deaths in construction work. (Photo courtesy of Constructioneer.)

At the hospital, it was discovered that the fall had numbed his body, but nothing was broken. The temporary paralysis was relieved with a shot and the warmth of the hospital. Late that afternoon he was back on the job.

The superintendent, looking serious, told David, "I'm docking you for the time you were gone. What's this shit, taking off half a day when we got a pour going?" David laughed, "Up yours, I'm reporting you for unsafe ladders." The super had the last word, "No more jobs for you over two feet."

The above incident reveals that men anticipate accidents and react accordingly, but they do not coddle themselves or each other. As will be explained later, they try not to show fear. As things turned out, David had another accident and was permanently injured, or so everyone (including David) thought. At age 27 he gave up construction work and went on permanent disability. He later moved to another city to seek another type of employment. Although the doctors told David he would always walk with a limp, he told himself they were "full of shit" and he embarked upon a program of weight lifting and exercising his leg. It gradually improved, and within two years he was strong enough to return to Buffalo and took his place once again in the construction industry. When he returned, he had renewed strength and vigor, and he quickly worked himself into a superintendent's position and soon was running his own jobs.

At Amherst, the pile-driving crew was sheet piling a sewer trench 25 feet deep. Steel I-beams were used to brace the pilings and were placed at the top of the sheets across the trench. The men had to go back and forth across the I-beams in order to get from one side of the trench to another. On the day of the accident, the hammer that drove the sheets into the ground (and eventually into rock 25 feet below) was hung up on the crane boom and the operator could not release it. It was dangerous, for if the weight dropped out of control, it could pull the crane over onto the men below. When the foreman saw the situation, he started running across one of the I-beams, slipped, and went down. Directly below him was a foundation for a manhole with reinforcing rods sticking up in the air.

The foreman luckily missed the steel rods, where he would have been impaled. Instead, he landed hard on a steel mat of criss-crossing reinforcing rods, broke six ribs, and punctured his lung. The problem was how to get him out of the trench. When the fire department ambulance arrived, he was wrapped in blankets and strapped into a stretcher. Then the men in the crew rigged a cable and hook to the stretcher and lifted him out with the crane. He never returned to the job.

CAVE-INS AND EARTH SLIDES

The federal Office of Safety and Health Administration considers "trench excavation" as one of the most dangerous jobs in construction work (*Construction Equipment* 1977:27). The reason is the frequency of cave-ins and death in this kind of work. I have never met a sewer line worker who did not have personal incidents to relate concerning cave-ins which caused serious accidents. Earth

Figure 8. A pile-driving rig and shell piles that have been driven into the ground at the Amherst sewage treatment plant. The hammer can be seen at the top of the rig that fits over the pile. (*Photo courtesy of John Kaiser, Roy Crogan & Son.*)

slides are also dangerous and fearsome. The following is a description of one that took place on one of the sewage treatment plant projects.

When we stripped the project of its topsoil, preparatory to the start of construction, we stockpiled the soil in three areas. Each pile contained 15,000 cubic yards of dirt. Later in the project, another contractor, working on the second phase of the plant (our contract was Phase 1), started a large excavation about 200 feet from our topsoil piles. In order to make room for other structures, we were asked to place one of our topsoil piles on top of another. This created a mound 30 feet in the air, containing 30,000 cubic yards of dirt and weighing 60 million pounds. I recall walking alongside the pile and having to strain my neck backwards to see the top of it. The combination of the excavation 200 feet away and the height and weight of that topsoil pile were the ingredients that produced the most massive earth slide I have ever witnessed.

On the day of the slide, I was inspecting the job, walking alongside the topsoil pile, and saw several bulldozers and trucks on top of it as more and more dirt was piled up. Later, sitting in the shanty, someone shouted over the two-way radio that there had been a cave-in. I grabbed a camera and ran out of the shanty with others, heading in the direction of the men I could see gathered near the topsoil pile.

When we arrived at the scene, the place where a short time before I had to look up, straining, to see the top of the pile, the topsoil was gone. It had sunk out of sight.

Where had it all gone? The weight of the dirt had pushed the clay below it down and toward the open excavation 200 feet away. This huge excavation was now filled with the clay that had been pushed up from below as the topsoil pile had dropped out of sight. The clay had slid and emerged in waves of solid, frozen-looking clay, jaggedly haphazard in all sorts of wierd shapes.

At the topsoil pile, two bulldozers were overturned and buried. Their tracks were off and their cabs jammed with dirt. One dump truck was pitched forward, its nose stuck into the ground. Another dump truck was on its side with only one half of it visible. In some places the earth was cracked open more than ten feet wide. There were cracks running in all directions.

I questioned the men in the area. No one could describe what happened. Everyone said it took place too fast. They all exclaimed disbelief at the sight. No one had ever seen such a massive dislodging of earth.

The fact that no one was hurt, killed, or buried was due to the fact that it had happened during lunch time. Had it occurred before or after, two bulldozer operators and two truck drivers would have been buried. The reactions of the men were very similar. Most felt humbled and subject to forces over which they had no control.

Figure 9. After the huge earth slide at the Amherst treatment plant project. Thirty thousand cubic yards of top soil, weighing 60 million pounds, just "disappeared" into the ground. (Photo courtesy of John Kaiser, Roy Crogan & Son.)

Figure 10. High steel men setting structural steel on a New York City skyscraper. (Photo courtesy of Constructioneer.)

There were other cave-ins on the sewer plant project, showing the danger and unpredictability and power of the earth when it is deeply dug into.

HIGH STEEL

Construction men who work on the skeleton of a building—carpenters, masons, and others—often must work on scaffolds at great heights. But they do not risk their lives each day the way high steel men do. Concrete workers, carpenters, and masons stand on platforms and scaffolds. A high steel worker must walk on beams that are only inches wide—from a perilous eight to a risky 24. They work exposed. The skin of the building is not yet up to protect them from the wind. Men able to work hundreds of feet in the air, walking on narrow beams of steel, are rare. Other men in construction respect their nerve and courage. Other construction men step in and out of danger. But the high steel worker faces danger in every minute, hour, and day while in the air connecting steel. He cannot make one false move. He must wrestle heavy steel sections swinging at the end of a crane whose unexpected movement could knock him from his perch. A sudden gust of wind could swing the steel into him and send him down to death or injury. Or he could simply lose his balance.

High steel men are dependent upon others, in part, for their safety. A crane operator explains (Terkel 1972:23):

> This is a boom crane. It goes anywhere from 80 to 240 feet. You're setting iron. Maybe you're picking 50-, 60-ton and maybe you have ironworkers up there 100, 110 feet. You have to be real careful that you don't bump one of these persons, where they would be apt to fall off. At the same time they're putting bolts in holes. If they wanted a half-inch, you have to be able to give them a half-inch. I mean, not an inch, not two inches. Those holes must line up exactly or they won't make their iron. And when you swing, you have to swing real smooth. You can't have your iron swinging back and forth, oscillating. If you do this, they'll refuse to work with you, because their life is at stake. They're working on beams, anywhere from maybe a foot wide to maybe five or six inches. These fellas walk across there. They have to trust you.

MISCELLANEOUS ACCIDENTS

Faulty Equipment Many accidents occur because of the failure of mechanical devices. I once saw a man get his eyes burned because an oil hose burst while he was operating a new backhoe. Hub Dillard commented about the same thing (Terkel 1972:24):

> A lot of stuff that comes out of the factory isn't exactly right. It's faulty. They don't know until it's used on the job. It's not just one person that's hurt. It's usually four or five. . . . There was 11 of them in an elevator downtown. The company that built that elevator, it was supposed to be foolproof. . . . It fell 12 floors and they were all hurt bad. Two of them had heart attacks when this was falling. There was one fella there that was completely paralyzed. He had 11 children. The only thing he could move was his eyes, that's all.

In 1978, 51 men were killed in a single accident in Willow Island, West Virginia, when an entire scaffolding system collapsed. The scaffolding ripped away from the sides of a cooling tower in a circular fashion, sending all 51 men working on it about 170 feet off the ground to their deaths in a heap of concrete, metal, and twisted debris. Federal officials of the Occupational Safety and Health Administration said it was the worst disaster they had seen, except for those related to mining. An investigation disclosed that the scaffolding system was faulty (*The Buffalo News*, April 29, 1978).

A Sharp Trowel Concrete finishing—what could happen in that trade that would be dangerous? Well, the trowel they use is razor sharp. One day Pete R.'s hand slipped into the side of his trowel and was slashed open from his small finger to his wrist. Iggy put a handkerchief over it. When Pete came into the shanty, I asked to see it and he took off the bandage. I found my neck tensing up. I sucked in my cheeks and tried to make my face rigid so it would not fall apart at the repulsive sight. I told Pete not to look. We took him to the hospital and waited an hour till the doctor came and asked to see it. When Pete took the handkerchief away, the blood squirted out in a looping arc, landing on the doctor's coat and leaving a stain from shoulder to sleeve. The doctor pushed the handkerchief back over the cut. It took more than an hour to fix Pete's hand with two rows of stitches and a splint to keep the skin together. But Pete was back working on the job the next day.

Two Nails One day, Iggy was helping to strip concrete forms when two nails at the end of a plank went into his hand. Being crossed on the board, they went into his hand the same way and were locked and crossed when they came out the other side. They entered the back of his hand between his thumb and index finger and came out inside his palm. There was no way to get the nails out without tearing away the flesh. The nails were at the end of a 12-foot board, and the weight of the board was exerting upward pressure on the nails and on his flesh, putting him in terrible pain. In spite of his pain, Iggy realized that the board was putting pressure on his hand and yelled for the board to be cut as close to his hand as possible. But the problem was how to get those crossed nails out. Again, in spite of the pain, he had the presence of mind to figure out how to do it. Iggy told one of the men to get a nail cutter and cut the heads of the nails off. Then they lifted the boards off the nails and were able to slide the nails out of his hand. He was taken to the hospital for a tetanus shot and bandaging. He was back on the job the next day.

This rapid spring-back is characteristic of construction workers who perform hard and dangerous jobs. They are contemptuous of those who "baby" themselves after an accident. They are equally contemptuous of men who endanger themselves stupidly.

Orange One lunch hour, a laborer named Orange was smoking a cigar near three propane tanks. He was in the intake chamber to the pump station, a large concrete structure that would receive all the town's sewage when the system was turned on. The men were using it as a storage area and a place for them to warm up—hence the propane tanks. The inevitable happened. There was a leak in one of the tanks and Orange, smoking too close to the tanks, set it on fire. Those who witnessed it said that Orange seemed to freeze, seemingly unaware of the danger, until someone yelled, "Get the fuck out of there, you dumb bastard!" Orange and everyone else in the area managed to get out in time. The three tanks blew and shot out of the chamber like naval shells, the steel jackets folding up like cardboard. The heat inside the structure was so intense that large chunks of concrete peeled off, making the walls and ceiling look as if they had been shelled. The piping inside the chamber melted. The men needled Orange for a long time, angry that he was so stupid as to smoke near propane tanks. They kept asking him when he was going to stop smoking his "dumb-ass cigars." He did stop for a while after the incident. But it was not long before he was smoking again, oblivious of the possibility of explosive gases in the area. He showed up at the farewell party for

Iggy a few months later, and someone remarked, "What, are you still around, Orange? I figured you must have blown yourself up by this time."

Danger on construction projects can come from the most unexpected sources. There was the time Tse Tse was working a concrete bucket behind a 12-yard concrete truck. The driver had backed the truck into position, jumped down from the cab, and started toward the rear to set the drum in motion. Suddenly, the truck began to roll backward. The gear had somehow jumped into reverse. Tse Tse managed to scramble out of the way without being crushed, but the truck continued to head for an open excavation where men were working 50 feet below. By some miracle, the concrete bucket got jammed underneath the truck and slowed it down long enough for the driver to jump back in and release the reverse gear. Had the truck gone over the edge, a grievous tragedy would have occurred.

Bob G. recalled how a concrete truck rolled over him. In fact, it rolled over his head. The doctor told him that it was a miracle that he was still alive. Fortunately, it was a very wet, rainy day, and the ground was extremely muddy. Bob's head was pushed into the ground, which was pliable enough so that he was not completely crushed. He suffered a concussion and many broken bones in his face, but after many months in the hospital he recovered and was able to get back into construction and become a superintendent. The men often kid him about being run over by a truck over his head, attributing his "dumbness" to that event. However, he is far from stupid—wild, yes, but not stupid.

Recognizing danger as a daily possibility in construction, the Occupational Safety and Health Administration monitors safety standards throughout the industry. OSHA inspectors visit construction projects and look for open stairwells, gases and chemicals that might explode, unsafe ladders or scaffolds, debris that could catch on fire, trenches that are unprotected or unsheathed, and the like. The OSHA manual sets forth recommendations for safety. Hard hats are a must—any worker who chooses not to wear one is a fool, since objects are always falling from buildings. Fire extinguishers are important because small fires are always breaking out from any number of causes. OSHA is particularly stringent on elevators and types of scaffolding that must meet safety standards. Yet, in spite of all precautions, there are a thousand ways in which an accident can take place each day. If a man is in construction, chances are high that he can be hurt.

THE NECESSITY OF APPEARING FEARLESS

Mike Cherry, discussing attitudes of high steel workers to danger, commented that it was important for a man in that kind of work not to show fear, especially to himself (Cherry 1974:149). This attitude sums up the general outlook of all construction workers. Dan L. recalled when he first started working as a surveyor. One of his first jobs was the Queenston–Lewiston Bridge over the Niagara River gorge, 368 feet above the river. He recalls the first time he had to walk out on the steel and hold a rod for the man on the "gun." Everyone was watching to see how he would react. They knew he was scared, but they wanted to see how he would

Figure 11. A view beneath a bridge we built on the downtown Buffalo road project. The bridge was 330 feet long and spanned two streets. The girders had to be spliced together. Each girder measured 110 feet long and weighed 45,000 pounds. (Photo courtesy of The Buffalo Courier-Express.)

behave before they decided whether to work with him. Dan said that the foreman of the bridge crew would watch men who were sent to erect the steel. The foreman would take one look at how a man walked out on the steel, and if he was dissatisfied he would remark, "You're not a bridge man. Go back to the hall." Jack Haas, in his study of high steel workers, noted:

> Workers who perceive physical danger develop mechanisms to control their reactions and the reactions of others. Individually and collectively they struggle to enhance the security of their situation. Symbolic or real threats bind workers together in an effort to protect themselves. Part of the defense, however, lies in controlling one's personal trepidations and insecurities and maintaining an appearance of fearlessness (Haas 1977:168).

The need to mask fear does not mean that men are foolhardy and will work under any conditions without complaining. One day, in January 1978, we were making a deep connection into an existing sewer on a downtown city job. The connection was

30 feet deep and involved an enormous excavation at an intersection. We had had several slides and cave-ins on that excavation, as well as broken electrical cables and telephone lines, until we finally managed to get the hole shored with large timbers, boards, and steel braces. The assistant superintendent on the project was responsible for setting up all excavations for the blocklayers building the manholes. The bricklayer foreman said that he was not going to permit the removal of any of the braces that were holding the timbers and boards, even though they were in the way of building the manhole. He said that the braces would have to be built into the manhole and burned out after the manhole was completed. The braces were hydraulic shoring braces that were rented from an equipment supplier and cost a good deal of money. The assistant superintendent taunted the bricklayer for showing fear over working in the hole without the braces. All of the other men in the room jumped on the assistant super for being willing to jeopardize the lives of the men over a matter of money. No one questioned the courage of the bricklayer foreman. Rather, he was seen as exercising good judgment in protecting the welfare of the men working for him.

THE RIGHT TO CONTROL THE WORK ENVIRONMENT FOR SAFETY

Jack Haas, in his study of the perception of danger among high steel workers, suggests that the perception of danger leads to similar processes and expectations in various occupational groups.

> The single characteristic they all share is their perception of danger; this perception produces a set of perspectives around the problem of danger that is rigorously and continuously enforced. The workers attempt to increase their control over their work environment and lessen the dangers. . . . This perspective emphasizes the ironworkers' commitment to increasing worker autonomy and thus a control over their environment. They strive to maintain control by collectively supporting individual and group decisions to judge for themselves safe and unsafe working conditions. They support the actions of fellow workers who decide whether or not to work in inclement weather. Fellow workers accept or reject the judgments of work superiors who may not give precedence to ironworkers' considerations and who could consequently pose a threat to their personal and collective security (Haas 1977:168).

The manhole incident is a good example of this type of mechanism. When the assistant superintendent seemed to be primarily interested in the cost of the braces, rather than the safety of the bricklayers, all the other men in the room, most of whom were not bricklayers, came to the support of the bricklayer foreman. The bricklayer foreman, on his part, was asserting control over the environment in which his men would work, with respect to whether or not it was safe.

Another illustration of the above principle occurred at the library project. Structural steel was being erected in the winter of 1975. There was a big push to get all the steel erected so the concrete floors could be started earlier in the spring and the brick started early enough to enclose the building before the following winter.

The architect's representative insisted there was no reason to halt erection of the structural steel, except when there were high winds which could be dangerous. One weekend, there was a considerable thaw together with a heavy rain, followed by very low temperatures. On Monday, when the ironworkers came in to work, the steward sent one of the apprentices up to check the steel for ice. The apprentice reported back that the steel was too slippery to work on. The architect's representative told the foreman to take some scrapers and chip the ice off. The foreman refused. The following conversation took place.

Architect's rep: "You mean to tell me these men came in to work and they're not going to because you refuse to chip off the ice?"
Ironworkers' foreman: "That's right."
Architect's rep: "Get the general contractor. Let's hear what he has to say."
Architect's rep to the general contractor: "I'm ordering you to work. You're aware of the schedule. We need the steel so we can get the floors so we can close in the building by fall."
General contractor's project manager: "I have no control over this matter."
Architect's rep: "Who decides whether they will leave?"
General contractor's project manager: "They do."

What the architect's representative did not understand in this case (and what was clear to everyone who has been a part of the culture of construction) was that when it comes to safety, no one dares to challenge the right of the men to determine for themselves what is safe and what is unsafe. It is interesting to note that in a "Worker's Bill of Health Rights" (Wallick 1972:109–114), the 15 articles included rights to protection from hazards and dangers; this protection was to be provided by others who controlled the work environment. Nowhere did the Bill of Health Rights state that the worker had the right to walk off the job or refuse to work if it was not safe. This is a right that construction workers enjoy because they have a large measure of control over their work environment.

IDENTIFICATION WITH THOSE WHO GET INJURED

The attitude of many construction workers toward the many accidents and deaths that take place is "there but for the grace of God go I." With such a high incidence of accidents, it is not surprising that most construction men feel it is only a matter of luck as to who will be the next victim. And the percentages are enormously high. On the District of Columbia's subway project, which was started in 1970, according to Clyde Farrar, Chief Inspector, one out of every three men working on the subway had been injured since work began (Scott 1974:254).

Dan L. was talking about Joe C. one day in December 1977. Joe was killed when a concrete truck went over him. Dan said he was Joe's foreman, and when he went to Joe's funeral he felt everyone looking at him as though asking why he was alive and Joe was dead. Dan said he could never forget the wide, innocent eyes of Joe's daughter looking up at him and really understanding nothing of what was going on. Dan said it made him think all the time about how his own daughter might have looked if he had been killed, instead of Joe.

The men identify with each other when one of them gets hurt. A pipe man was struck by a section of 84-inch pipe that weighed 12 tons. It swung around while the men were trying to put it in a trench and hit the worker in the chest. He was described as "a good worker, a family man . . . a man who spoke with an Irish brogue and was liked by everyone . . . a guy who smiled a lot and who had the gift of gab." When they got the injured man to the hospital, his work mates were told that he had some crushed ribs and would be all right. What no one realized at the time was that he had lost consciousness while on the way to the hospital and had had a stroke. For a short time the oxygen supply to his brain had been cut off. As a result of the stroke, the man suffered the loss of the use of his right arm and considerable loss of speech.

The man hired a lawyer and sued for damages. The case involved the amount that he would collect. During the trial, the lawyer for the insurance company kept badgering the witnesses, openly calling them "dummies" and "stupid," implying that they were uneducated workers who could only lay pipe and were not intelligent enough to recall anything accurately. The lawyer was trying to discredit the witnesses by catching them in contradictions, thus imputing their veracity. Some of the men did become confused and contradicted themselves, and the lawyer succeeded in getting their testimony thrown out.

Then the lawyer focused on the man who was injured. He demanded that the man raise his arm, implying that he was faking it and could move his arm. The man painfully tried to raise it with the lawyer shouting at him that he was a fake. One of the workers who was there as a witness for the man jumped the rail where he was sitting and started for the lawyer. The others, while restraining the man, started shouting obscenities at the lawyer, threatening him with bodily harm when they got him out of the courtroom.

Dan commented about the court scene:

> we all felt that we were being insulted and attacked as a body. You could see that the man was paralyzed. Also, when he was questioned he had difficulty getting the words out. It was as if the lawyer was mocking his disability. And this guy was going to have to live with his disability for the rest of his life. I tell you the truth, I took it personally even though he was not a member of my family. I could have killed that lawyer. We all could. What happened could have happened to any one of us.

PHILOSOPHIZING

One Friday in January 1978, a bunch of us were sitting around at lunch time, swapping stories about accidents and deaths on construction jobs. Out of the eight men in the room, seven could recall from personal experience a serious accident or death they had witnessed or been involved in. Paul G. told how he was toppled into an excavation and how his arm struck a rock and was shattered. It was hanging limp when they pulled him out and started to swell from the bleeding. It was repaired with two cobalt plates and eight screws through the bones in his arm. Kenny E. told about his brother Mike who was buried in a trench cave-in and had suffered internal injuries. Kenny said as they were digging him out they could see blood coming out of his nose and ears. One of the men was working on the sewer job where five men were drowned one night on the night shift, when heavy rains caused a flash flood that filled the trench in which the men were laying pipe. None of the men questioned the fact that they had to work under dangerous conditions. They all complained about how much safety practices were ignored by everyone, including themselves. But they seemed to take a fatalistic attitude. Some dwelled on the flash flood that had drowned five men. With hindsight, it seemed that the men might have refused to work, surmising that the heavy rains would cause a flash flood which could threaten their lives. It seemed farfetched that anyone could have foreseen those possibilities and acted upon them.

We were searching for an answer as to the "why" in various accidents in construction. Why are some accidents fatal and others not? Why are certain men killed and others spared without a scrape? Some agreed that accidents are often mistakes. But when we returned to the drowning incident, it was agreed the flash flood was not the result of a mistake. Finally, after a long silence, someone said, "What's the use of looking for a 'why'? An accident is something that happens. There is no 'why.' It just is. That's all there is." That blunt realism epitomizes construction work culture.

7/Job satisfaction in the culture of construction workers

INTRODUCTION

Job satisfaction is the degree of positive attitude an individual has toward his work. Since Hoppock's study on job satisfaction in the thirties (1935), a substantial amount of research has been conducted on this topic. Job satisfaction is typically measured by interviews or questionnaires. Workers are asked to state the degree to which they like or dislike various aspects of their work roles. The degree to which a person is satisfied with his job is inferred from his verbal responses. Unfortunately, there has been little standardization of job satisfaction measures. Some investigators tailor their research instruments for the population they are studying. Thus, more indirect methods have been developed for evaluating job satisfaction (Weschler and Bernberg 1950; Weitz and Nuckols 1953). Job satisfaction scales were devised which were thought to measure job satisfaction irrespective of occupational population. Examples of these are the Brayfield-Rothe job satisfaction scale (Brayfield and Rothe 1951); the Kerr Tear Ballot (Kerr 1958); and the Job Description Index (Smith 1963).

One problem with job satisfaction research is that people have different reasons for their satisfaction or dissatisfaction with their jobs. Responses are different because people have different supervisors, coworkers, and job duties. They work for different companies and organizations.

My discussion will focus on factors isolated by researchers as significant sources of job satisfaction. I will then relate these factors to construction workers.

SUPERVISION

Herzberg and his associates (1967) compiled data in which workers were asked what made them satisfied or dissatisfied with their jobs. Supervision was mentioned more frequently than security, job content, company and management, working conditions, and opportunity for advancement and wages. The only aspect of the job mentioned more frequently was relationships with co-workers.

Much research on supervision is concerned with the degree to which supervisors consider the desires of their subordinates. One study by the Survey Research Center at the University of Michigan (Katz, Maccoby, Gurin, and Floor 1951) contrasted employee-oriented and production-oriented supervision. An employee-oriented supervisor established supportive relationships with his subordinates, took a personal interest in them, and was understanding when mistakes were made. Production-oriented supervisors viewed their subordinates as "people to get the work done" and were concerned with achieving high levels of production. There is evidence that satisfaction of subordinates is related to the employee orientation of their superiors. Supervisor behavior which is described as employee-oriented includes the following elements (Likert 1961:16–17):

Recommends promotion, transfers, and pay increases.
Informs workers on what is happening in the company.
Keeps workers posted on how well they are doing.
Hears complaints and grievances.
Thinks of employees as human beings rather than as persons to get the work done.
Will go to bat or stand up for the workers.
Usually pulls for the employees or for both the employees and the company, rather than for himself or for the company only.
Takes an interest in the employee and understands his problems.
Is part of the work group and takes an interest in people who are in the group.
Likes to get ideas from the people in the work group.

A number of studies were conducted on the relationship between supervisor consideration and absences, turnover, and grievances (Fleishman, Harris, and Burtt 1955; Fleishman and Harris 1962). The findings showed that the supervisor's consideration of subordinates results in job satisfaction as reflected in low turnover rates, absences, and grievances. Although this is probably the most likely explanation, the use of the correlational rather than experimental method leaves open another explanation. Possibly the direction of causality is opposite. Supervisors display a greater degree of consideration for subordinates they perceive to be satisfied and responsive. Subordinates who are critical and dissatisfied are less likely to be the recipients of warmth and personal support.

Using the ideas on the relationship between supervision and job satisfaction, let us examine the construction worker. Absenteeism, turnover, and grievances are almost nonexistent among construction workers. During three years of research, I was aware of five cases where men were fired because of chronic absenteeism. They were all new to construction. On my present road project, it is rare for anyone to be absent. Every man is accounted for at the end of each day as we plan for the next day. If anyone does not show up and does not call, a call goes to the man's house. If someone is absent from a crew, some shifting of men must take place. Every absence must be explained. If a man must take care of personal business and informs his foreman, he will usually not be docked. Grievances between the men or between the men and their foremen are usually settled without management's involvement. As project manager, I was only aware of one case on the road project where a man expressed a grievance about the way he was treated.

Consideration for subordinates by supervisors was high among the men I studied. Many of these men know each other and their families through kinship or friendship. After work, they participate in activities together—drinking, bowling, eating out, going to hockey games. Foremen and superintendents are considered part of the work group. If a man gets sick or is hurt, he is visited by the men and his status reported to others. Superintendents and foremen buy coffee or beer for their crews. One foreman, when he talks to his men, puts his hand on their shoulders or holds them by the arm or the back of the neck, exhibiting a feeling of warmth and human consideration for his men as persons.

It may sound like a contradiction, but in construction supervision is the key to the success of a project; yet, it is loose rather than strict. Supervision is responsible for organizing the necessary materials, equipment, and labor. Construction is unlike manufacturing, where materials, men, and machines are assembled at specific work areas. A construction project is a large site where men, machines, tools, and materials must be constantly moved from one work area to another. It is the job of supervision to plan these moves in the proper sequence. Supervision must know when and which trades must be alerted to start their operations, while being aware of others' schedules.

Supervision is loose in that the superintendent does not direct the work. He orders the foreman of a crew to a particular station and tells him when he needs the work completed. The foreman directs the work and works along with the men. The foreman plays both an authoritarian role and a cooperative one, since he participates with the men in performing the work.

Supervision in construction is democratic in style, relying on the men. It is not possible to subject the men to a strict routine. Supervision cannot be too authoritarian since it must ultimately depend on the journeyman to produce the work, and authoritarianism in construction (and perhaps elsewhere) inevitably leads to either resistance or walk-outs. Neither can supervision be laissez-faire, because, after all, the job must be completed. The democratic style of supervision is inevitable in a work situation where craftsmen are the main producers and where they control the work process.

Field supervision—foreman or superintendent—is the highest rank a journeyman can aspire to in construction. Thus, there is never a wide gulf between the men and their supervisors. The foreman of today was the worker of yesterday. For example, on our road project, when we expanded the number of operations on our waterline installation, we had to split up the old crews, make new ones, and appoint new foremen. We needed crews to pressure-test and chlorinate the lines; to make new house connections from the new main; to install new fire hydrants and street valves; and so forth. Men who had never been foremen found themselves suddenly in charge of an operation and a crew of men. Some were scared. It took about a month for them to become sure of themselves. Soon they began to demand things—tools, equipment, and materials. They became angry if their crews were delayed. They were kidded by the other men in their crews, and their responses were false threats to give their hecklers "dirty jobs in the hole." The kidding helped relax the new foremen. It meant their men were backing them. These new foremen were effective because they shared with their subordinates participation in the work process. They could not be authoritarian. They would have looked ridiculous if they tried. Their situation called for democratic leadership, and it paid off in the men's acceptance of their leadership.

Although an official has the right to command and his subordinates are obliged to obey, a superior's authority over subordinates extends only as far as they voluntarily permit to be governed by his directives. In construction, actual authority is not granted by the formal organizational chart. It must be established in the course of social interaction that is appropriate to the situation. Supervision which is democratic and shows consideration for subordinates is the most effective. This is not only true in construction but in other fields as well. A study of 24 clerical sections in an insurance company showed that productivity was higher when employees were given more freedom to do the work in their own way (Katz, MacCoby, and Morse 1950:17–29). Other studies have shown that disciplinarian supervisors are less effective than more liberal ones (Roethlisberger and Dickson 1939:452–453).

PARTICIPATION IN DECISION MAKING

One of the basic assumptions of those who research job satisfaction is that persons obtain satisfaction from influencing decisions and controlling their work environment. Evidence in the literature supports this view. Baumgartel (1956) studied scientist attitudes in 18 laboratories and found significantly more positive attitudes where they were permitted greater participation in work decisions. Jacobson (1951) found the same thing in an automobile plant. Wickert (1951) studied telephone operators and service personnel. He found major differences in the degree to which those employees who remained with the company were able to influence their job conditions as compared with those who left. Ross and Zander (1957), in a study of 2680 female workers in a large company, substantiated the Wickert study.

Morse and Reimer (1956) carried out research on job satisfaction among four clerical operations of a large insurance company. Two programs were set up. One was an autonomy program and the other a hierarchically controlled program. Job satisfaction was measured before and after an experimental year. There was an increase in satisfaction under the autonomy program and a decrease in the hierarchically controlled one. Two other studies found that workers display higher job satisfaction when they are able to decide about products, goal-planning, division of labor, and work assignments to group members (French, Israel, and As 1960; Kay, French, and Meyer 1962).

To sum up, there is fairly clear-cut evidence that people who are satisfied with their jobs tend to report that they have greater opportunity to influence decisions which affect their work and job environment.

Relating the above generalization to construction workers, we can examine the activities of a typical work crew installing waterline pipe on the downtown road project. This crew had the task of installing approximately 5000 feet of water main pipe.

Each morning, the crew would decide what fittings they needed, based on where they had left off the previous day. The foreman would ascertain if he needed a pavement breaker and, if so, would request a machine and an operator for a particular location. The foreman would have to make sure the public did not park in the work area. Someone would be assigned to come in early and keep the cars away.

There were four men in the work crew: an operating engineer, who ran a backhoe machine which dug the pipe trench; a foreman, who directed the operation from the top of the trench, signaling and talking to the machine operator on top and the men below; and two laborers in the trench, where they set the pipe, bolted up the fittings, and did whatever hand digging was necessary. One of the laborers also served as a flagman, directing traffic when the machine had to move back and forth into traffic backfilling the trench. The crew also put up movable construction signs each time they moved to a work area. These signs warned the traveling public that construction was going on in that area.

The men worked as a team, collectively solving their daily problems. Their greatest difficulty was at each street intersection. At those places they encountered existing underground utilities—gas, electric, telephone, sewer, water, telegraph, fire alarm, and others. Sometimes, the waterline would have to be installed in a snake-like configuration over and under these various utilities. At times, utility companies would have to be called to remove and relocate their lines so the waterline could get through. If the utility company had to be located, the crew notified the field office. The foreman and the crew would decide amongst themselves whether they could get through an area. If additional waterline fittings were needed, they would call the field office and tell the clerk to order them. The crew would have to look at the plans to see what provisions would have to be made for house services, fire hydrants, and street valves, and they would install fittings in the pipe to receive these various items. There were times when the crew would have to call upon the project manager or superintendent to make a decision. Occasionally, the state engineer needed to be consulted if a decision involved a design change or a monetary change order. It was up to the waterline crew to bring these problems and questions to the attention of the contracting company's supervisors or the state inspectors on their operation.

It was rare when crew or foreman decisions on the waterline main were overruled. It only happened once, and that involved a matter of safety. The men were not using the steel box that served as a safety shield because they could get more production without it. They were ordered to use it. If any of them got hurt, the project manager or superintendent could go to jail, under new federal safety codes.

George Strauss sums up the construction worker in this manner: "If we take the building tradesman as the classic example of the craftsman, we find that he behaves in many ways like a professional. . . . Craftsmen like to decide how they will do their work" (1963:16).

MEMBERSHIP IN AN INTEGRATED WORK GROUP

A considerable body of research exists which asserts there is a strong relationship between greater job satisfaction and work based on integrated work groups. Herzberg and his associates, referring to 44 articles on job satisfaction, states: "These findings indicate that the cohesiveness of the work group is an important factor in determining the morale of workers" (1957:132).

There are two aspects to integrated work groups. The first has to do with the techniques and technology of the work itself. Since construction tasks usually involve teams, crews, and gangs, construction work inherently fosters cohesive work groups. The other aspect of integrated work teams is the social one. On most construction projects, a community social structure is usually built along with the physical structure. Friendships are formed and grow between men who know each other from previous jobs or meet in the same crew for the first time. There is a social system of leadership and groups based on the trades and the social roles on the project. Informal leaders emerge through reputations based on skills, knowledge, and personality. Channels of communication are established for work, social, and other matters. Social interaction takes place at work, during lunch time, and after work at the local bar or bowling alley. All of this activity builds group feeling, a sense of "us" and "they."

Within the total community of men on the project, there also develop small groups and cliques, usually based on trade affiliations but also on previous friendships. Hub Dillard talks about the cliques based on trade (Terkel 1972:24):

> You're tense and most everybody'd stop and have a beer or a shot. They'd have a few drinks and then they'd go home. They have a clique, like everybody has. Your ironworkers, they go to one tavern. Maybe the operators go to another one. The carpenters go to another place. They build buildings and tear 'em down in the tavern.

On one of the sewer projects, the pile drivers had a clique and a special reputation. They were the wild bunch. They did crazy things, engaged in horseplay, and were always fighting with each other, both verbally and physically. One time they toppled a crane onto a small steel building on the job site. They thought it was hilarious, seeing men scatter in all directions as the crane slowly started to tip and then came crashing down on the empty building. Another time they lost their foreman, who fell into a trench, cracked his ribs, and never returned to the job.

The pile drivers drank in the same tavern where several other groups drank. They had their own place at the bar and made plenty of noise. Sometimes, they gathered at one corner of the room and shoved several tables together. They welcomed no outsiders and none showed any inclination to join them. They were thought of as rough, sloppy, and uncaring. They usually finished their work, but were held in low regard by the other trades because they took their time and delayed other crews.

On the other sewer project, we had some pile driving work to do, and some of the same men were part of the crew, but they had a different foreman. The group had an entirely different personality. They worked efficiently, had no mishaps, and

Figure 12. The pile-driving rig and crew at the Amherst sewage treatment plant. The plant was capable of handling sewage for 1,000,000 people. The pump station was built on solid rock–glacial till. But all other buildings were built on varved clay, or "gumbo," and the structures had to be supported by shell piles, which were driven into the ground until we hit solid rock. (Photo courtesy of John Kaiser, Roy Crogan & Son.)

were either ahead of the other trades or on schedule. They gained the respect of all the other trades. I asked one of the members who had been in both crews why he thought they were so different on the two jobs. He answered, "I guess every crew has its own ways. To tell the truth, I enjoy both. Don't ask me why."

In addition to integrated groups that form on the job, there are crews that go from job to job. The two sewer plant projects overlapped and we had 30,000 cubic yards of concrete to pour on the two jobs. The work was not continuous, and we hired concrete workers as we needed them. It worked out, however, that the concrete crew could move as a body from one project to the other, meeting the needs of both in this intermittent manner. The crew had been in existence a long time. We had used them at the library. They worked like a machine. Each man knew his job as well as the moves of the others in the crew. George Strauss talks about this kind of arrangement in construction (1958:65):

> In some cases (particularly among ironworkers) "crews" went from job to job together—even from one city to another. Often they took their own foremen with them. If the boss fired one of the crew, the rest quit too.

The effect of integrated work groups on output and productivity has been given much study. An investigation of construction workers in Chicago demonstrated that both productivity and job satisfaction went up when a system of letting the men choose their own work partners was introduced (Van Zelst 1952:175–185). Applying the sociometric principles of Moreno (1937) to 38 carpenters and 26 bricklayers, work teams were regrouped into mutual choice crews and researched over an 11-month period. Controlling for other factors that might have affected output, Van Zelst found that it was clear that group output was traceable to the system of letting the men choose their own work teams. There was also evidence that job satisfaction increased. There was a drop in the number of workers who left the job during the period of the experiment (1952:183). Van Zelst concluded (1952:184–185):

> It must be noted, however, that the building trades with their "buddy-work-teams" are especially suited for a sociometric regrouping. . . . The end result in this study has been a happier, more productive worker, who has given management a 5% savings in total production cost.

One of the workers summed up his feelings (Van Zelst 1952:183):

> Seems as though everything flows a lot smoother. It makes you feel more comfortable working—and I don't waste any time bickering about who's going to do what and how. We just seem to go ahead and do it. The work's a lot more interesting too when you've got your buddy working with you. You certainly like it a lot better anyway.

The picture of the construction worker's situation contrasts with that of the factory worker. Work on the assembly line leaves no time to socialize because "the line-tender must do all the work that the endless belt brings before him" (Chinoy 1955:71–72). Job dissatisfaction is reflected in high rates of absenteeism and turnover (Walker and Guest 1952:62). Many factory workers perform their tasks alone (Levison 1974:61):

> Another aspect of blue-collar work is loneliness. Many working-class jobs are very consciously designed to keep people apart. In many factories the noise level alone prevents any conversation beyond a shouted remark or two. In others the work stations are far apart or the job too difficult to permit any real communication. Rules against talking are also common, either written down on paper or delivered by the supervisor. The attitude is that if you're talking you're probably not doing the job. Only where the work is necessarily done in teams, *like construction work*, {italics added} is there any opportunity for real contact.

Mining is another occupation that stresses work in teams and has given the men job satisfaction. Friedmann and Havighurst discuss it (1954:176).

> Coal miners had a very personal sense of being pitted against their environment and expressed feelings of accomplishment and pride at having conquered it. . . . The whole gang is pitted against the mine. Victory is dependent upon each man's holding up his end. Victory is not only the sense of having cheated death . . . it is also the sense of having achieved life through one's own efforts and those of one's fellow workers.

Figure 13. Two men working together with a plate tamper. The machine is driven by gasoline motors and used to compact dirt and fill. (Photo courtesy of Constructioneer.)

Another example of the effect of integrated work groups comes from the British coal mines. In British mines, coal was extracted by pairs of miners who shared between them all necessary tasks (Trist and Bamforth 1951:3–38). The men in these pairs usually chose each other as work partners. They were a close, integrated team, sharing a dangerous and difficult job. There was some division of labor, but each man carried out all the tasks. The system was inefficient and costly and not easily mechanized. Then, machinery and mechanical conveyors were introduced. There was separation of tasks and specialization of labor. The former short wall work areas were replaced by long walls. Instead of two-man groups, large gangs of men worked in rotation, each group carrying out only one task.

This change should have resulted in a large increase in output. It did not. Coal production hardly rose. Instead, absenteeism increased. Reported illness among the workers rose and tension levels became high. The Tavistock Institute of Human Relations investigated the reasons for the rise in worker dissatisfaction.

The Tavistock Institute found the breakup of the two-man teams had two main results. First, the men lost their sense of identification with the job. They no longer were close to the whole task to care about how much coal they produced. Second, rivalries between gangs interfered with coordination of the various stages of the job, a coordination easily achieved by the former two-man teams. One task was exceptionally dirty and considered "bad work." The group assigned to this task had low social status and responded by slowing down the whole mine's operations. Miners with previously high morale in the face of hazardous mine conditions now found themselves without the support of the closely knit team they had worked with before. Trist and Bamforth reported a consequent real drop in job satisfaction (1951:38).

Summarizing, we can state that most people are more satisfied when they work as members of a group than in isolation. Most workers prefer jobs that permit interaction and are more likely to quit jobs that prevent contact with others. Congenial peer relationships are cited by workers as among the major characteristics of good jobs. On the other hand, "low job satisfaction is related to . . . lack of opportunity to interact with coworkers" (Kasl 1974:176). Construction workers are fortunate in that the technology of their work demands that tasks be performed with teams and crews. Human sociability is not a substitute for satisfaction lacking in the job, but it is an intrinsic part of the work environment. Construction work requires cooperation within crews and between trades. Thus, the cultural pattern of behavior as well as thought engenders feelings of being part of integrated groups.

MEMBERSHIP OR BELONGING TO AN OCCUPATIONAL COMMUNITY

There is satisfaction from the feeling of belonging to a group in which one shares ideas and beliefs and where members engage in similar behavior. The notion of occupational community is based on the idea that men who work in the same trade, craft, or occupation share experiences and a way of life, so that as a group they have a culture. Community is used here in relation to a limited number of people. It is the number of people that are in face-to-face contact on a construction project during the life of the project. Entry into this community is through membership in a trade or craft, gained by the acquisition of the requisite skill. Being part of a craft implies that one is a certified member of the group, sharing the standing of other members. This gives self-esteem and prestige which one shares with others.

E. C. Hughes posed a number of questions for investigators of occupational communities (1959:25):

> To what extent do persons of a given occupation "live" together and develop a culture which has its subjective aspects in the personality?
>
> Do persons find an area for the satisfaction of their wishes in the association which they have with their colleagues, competitors and fellow servants?
>
> What part does one's occupation play in giving him his "life organization"?

An occupational community represents a particular relationship between work and nonwork lives which, in its ideal form, is rare in modern societies. It implies that members of an occupational community are affected by their work in such a way that their nonwork lives are permeated by their work relationships, with their attendant value systems. Adam Curle, in his anthropological look at "work," points out that societies of the nonindustrial type have no notion of the separation of work and nonwork lives (1949:41–47). Sometimes, they have no word for "work." In our society, our work lives are isolated from family, political, and religious roles. Many people believe they do not start to live for themselves until they leave work—like the packing-house worker who says that he is anxious to get home from work so he can "try to accomplish something for that day" (Blum 1953:96). Gerstl comments (1961:37) that "work offers few positive satisfactions and is not a central life interest for the majority of industrial workers."

Differing from this condition are the attitudes and life-styles of many craftsmen, described here by C. Wright Mills (1953:223):

> The craftsman's work is the mainspring of the only life he knows; he does not flee from work into a separate sphere of leisure; he brings to his non-working hours the values and qualities developed and employed in his working time. His idle conversation is shop talk; his friends follow the same lines of work as he, and share a kinship of feeling and thought.

A good example of Mills' craftsman is the stone mason who even daydreams about his work (Terkel 1972:xlvi): "I daydream all the time, most times it's on stone. All my dreams, it seems like it's got to have a piece of rock mixed in it."

Occupational communities typically can be broken down into four components: 1) self-image; 2) accepting evaluations only from peers; 3) integration of work and nonwork lives; and 4) sharing problems and experiences.

Self-image A self-image is the way a person views himself. This self-perception is not accidental. It is based on the social role of the individual which receives support and confirmation from others. When a man's self-image is centered on his occupational role, those who give him support and confirmation are his work mates. Robert Park supports this view (1968:94):

> The conceptions which men form of themselves seem to depend upon their vocations, and in general upon the role which they seek to play in the communities and social groups in which they live, as well as upon the recognition and status which society accords them in these roles.

People who value their work and get satisfaction from it are likely to take their self-image from their occupation. Construction workers have a positive self-image based on occupation. There are indicators which attest to high prestige. They are among the highest paid, if not actually the highest, of all blue-collar workers. They are given recognition and autonomy in their work based on their skills. Some researchers have even likened construction craftsmen to professionals (Stinchcombe 1959:168–187). They are given the freedom in their work to independently organize and perform their skills that is similar to the autonomy of professionals. Construction workers have been referred to as the "aristocrats of labor" (LeMasters 1975). This reflects the prestige and self-image they enjoy. George Strauss talks about the similarity between construction workers and professionals (1958:65):

> The construction workers studied were intensely proud of their craft, showed strong social unity, and were independent and self-assertive. Their identification with their craft approximated that of a doctor or lawyer.

Construction workers all have some title that identifies their membership in a craft or trade. It is a title which certifies membership and identity. It gives those who bear this title prestige and a sense of belonging. If a man is on a construction project, he is known not only by name but also by trade. Even if someone does not know his name he will be called "electrician" or "bricklayer" or whatever. His clothing and his tools will identify him, as well as the kind of work he is performing. The importance of occupational title to self-image is explained by Becker and Carper (1956:344):

> Kinds of work tend to be named, to become well-defined occupations and an important part of a person's work-based identity grows out of his relationship to his occupational title. . . . Occupational titles imply a great deal about the characteristics of their bearers and these meanings are often systematized into elaborate ideologies which itemize the qualities, interests and capabilities of those so identified.

The positive self-image of construction workers is revealed in two other ways. One is the frequency with which they take their families past projects they have worked on and point with pride to their participation in the job. The other is their encouragement to their children to enter the industry. In contrast to industrial workers, the building tradesmen researched were, in most cases, anxious that their children follow in their footsteps. On every project from which data for this study were drawn, both fathers and sons were working, and many children of tradesmen working on the job were apprentices.

Accepting Evaluation Only from Group Members Just as doctors and lawyers believe that only their colleagues are competent to judge them, construction workers stress that only fellow tradesmen can judge their work performance. A remark made frequently by construction men to inspecting authorities and owners is, "You can inspect the work when it is finished, but you cannot tell me where, when, and how to perform my job." At one of our sewage treatment plant projects, we had a very competent resident engineer who performed his administrative duties well. However, he engendered great hostility when he tried to interfere with the work process and tell the men how to perform their work. After many showdowns, the resident was forced to confine his review of the work to final results and allow the craftsmen to judge their performance through their own craft foremen and skilled journeymen. The insistence on judgment of performance only by one's peers is explained by Gouldner (1960:468):

> Continued standing as a competent professional often cannot be validated by members of his employing organization since they are not knowledgeable enough about it. For this reason, the expert is more likely than others to esteem the good opinion of . . . peers.

In an occupational community situation, peers not only evaluate the work skills of colleagues, but their habits and personality traits as well (Caplow 1954:127):

> Those which are disapproved are the ones which interfere with the smooth functioning of the working group: pugnacity, carelessness, taciturnity, nervousness, dishonesty, self-pity, and the like. In a sense, the rules of comportment which are applied are extensions of the tacit rules encountered in the "normal" family.

Convergence of Work and Nonwork Life Members of an occupational community tend to prefer friendships with people who do the same type of work. This means more than merely being friendly with people at work. It means spending time outside working hours with work mates. Work mates predominate as best friends. In their study of printers, Lipset, Trow, and Coleman make this point (1956:70):

> Large numbers of printers spend a considerable amount of their leisure time with other printers. In interviews many printers reported that their best friends are other printers, that they regularly visit the homes of other printers, that they often meet in bars, go fishing together, or see each other in various places before and after work.

I tested the question of friendship by taking a sampling from one of the sewer plant projects. I asked 11 men I knew well to tell me how many "close" friends they had who were construction workers. "Close friends" were defined for them as men they knew more than five years, those with whom they exchanged home visits, and those with whom they shared after-work activities like bowling, hunting, fishing, or drinking. I told them to consider close friends as those who had all three characteristics, not just one. I also asked them to list nonconstruction friends defined in the same manner.

The table which follows shows the results.

Subject	*Construction Worker Friends*	*Nonconstruction Worker Friends*
E.K.	4	1
D.C.	3	0
B.M.	1	1
I.P.	3	2
K.O.	0	0
P.R.	2	1
S.F.	1	0
B.L.	0	2
P.C.	3	1
C.C.	2	0
B.H.	2	1

The results above confirm the tendency of construction workers to choose friends and social contacts from among their occupational group.

It is by no means common for men to have friends from their occupation. "Occupational communities rarely exist among urban factory workers" (Blauner 1966:483). Many workers attempt to "insulate" their nonwork lives from any work influence. Goldthorpe and his associates found that the majority of men in their study sample neither had relationships of any depth with their work mates, nor were they interested in finding friends from their work place (1968:56):

> Taking the sample as a whole, only around one in four of our affluent workers could be said to have a "close friend" among his mates in the sense of someone with whom he would actually plan to meet for out-of-work social activities.

Sharing of Problems and Experiences Danger and discomfort in construction seem to be common elements in construction workers' lives. The men enjoy swapping stories about experiences that include these elements. They often relate such incidents in a humorous manner, with several men joining in with their own variation of the type of story.

One favorite source of tales concerns difficulties with construction equipment. Company policy through the years has been to keep company-owned equipment as long as possible, even if it is obsolete and falling apart. Arguments about less down time and more efficiency with more modern equipment have not dissuaded the company owners from their policy. Thus, the men have been given construction machines to use that were so bad they became famous or were given nicknames. One of these machines was a small bulldozer, the subject of stories by many of the men. One of the stories told one day at lunch went like this:

> We're in ________ on the state job and I've got the mini-beast, you know the one, the Case 190, and we're doing some clearing and grubbing. It's January and we got to jump it every morning to start it. One morning we're pumping gas into her and it's leaking all over the carburetor. Timmy's putting on jumper cables and don't you know he gets them mixed up, causes a spark and starts the machine on fire. I'm across the field in a car with the inspector and I see Timmy running across the field. I also see the fire so I grab a fire extinguisher and start running. When I get there I push the spray nozzle and my prick could have pissed more of a spray. Then we run and get the water truck and drive it hard across that field. It was so bumpy our hard hats were hitting the top of the cab. We get there and open the hoses and the fuckin' water truck is empty. So we get a second truck over and finally get the fire out. We all feel like heroes. This one got a singe here, that one got a burn there but no one's burnt bad.
>
> That Friday, I go to the office to turn in my time sheets and who do I run into but K________, the big boss. He says to me, "Hey, there's Smokey the Bear. I heard you put out a fire on the Case 190. You dumb bastard, why didn't you let that piece of junk burn?"
>
> I told him if I ever see his house on fire I'm goin' to let it burn. Was I pissed. Here I thought I'm a hero and instead I get my ass chewed out.

During the same lunch time, another man told about a crane that would not turn to the right. Every time the men used it they would tie a rope to the boom and three men would pull the rope to turn the crane. One day, the boss came out to the job and saw what the men were doing. He commented, in all seriousness, "Boy, that crane sure works fine."

The men also like to tell stories about goof-ups, embarrassing moments, fooling inspectors and funny situations. One man told a story about a concrete pour on a thruway job. The men were standing around a long time waiting for the concrete truck to show up. When it finally did come, one of the apprentices said, "I've got to take a shit." The foreman got angry and asked him why "in the hell he didn't take his shit when they were all standing around." He told the apprentice he'd just have to wait until the concrete was poured, that they couldn't spare anyone. After the concrete was in place, the foreman said, "Now you can go take your shit." The apprentice said, "I don't have to, I shit my pants already."

Another man told about an incident on the same job. They had an old-timer on the job who they used as a flagman. They stationed him a half mile from the construction work to wave off the traffic to the other side of the road. The flagman was too far away to see what was going on with the construction. He would bring a chair, his lunch bucket, and something to read when there was no traffic. At the end of the day's work or at night if they worked overtime, the men on the construction gang would drive to his station and pick him up. One night the men went back to their hotel and forgot to pick up the old-timer. They found him the next morning, in his same spot, still flagging traffic. He was so used to their working overtime, he had just assumed they were working a 24-hour shift.

These stories help to cement the feelings of togetherness among the men. Although many of them are humorous, they relate to problems and difficulties which all can share. It fosters the sense of group membership, often articulates values and ideas which they unconsciously share, and gives everyone who participates feelings of comradeship in the exchange of jokes, stories, and even the remembering of tragedies. Here is a conversation that took place in a round robin fashion one day after work, when the men stayed for a long time, "shooting the shit":

Tony Z. started it off (Tony was a handsome Italian, with a remarkable wit and a delivery not unlike Bob Hope's, who was always a delight in any conversation, and who was also an excellent earth-grading foreman who had been in the industry for 20 years): Tony said that lately everyone was treating him so nice, he decided to come into the field office to see Sal and Danny so he could "get humble" again. Tony said when he was previously on the Union Road job with Sal, he broke the record—he got laid off seven times in one week. Sal would lay him off at 4:30 and then would call his wife in the morning and ask where the hell he was, he was supposed to be working. He said one day he got laid off once in the morning and once in the afternoon.

"A Sewer Rat" (Tony, like other construction workers, had no fear of or reverence for his superiors, so he insulted his foreman and superintendent as freely as he did others): Tony said Sal was a natural sewer rat. He said that every so often Bobby would say to Sal, "Come on, let's take a walk," which meant a walk through a sewer.

"Stealing" (one of Tony's favorite subjects): Tony said some days he gets up in the morning and sneaks an extra pair of socks and puts them in his coat pocket, "just to start the day off right." Then Danny mentioned another guy who was famous for his stealing. On Danny's birthday this guy stole a box of cigars for

him. Then he found out it was the wrong brand, so he took it back, got a refund, and on the way out, stole the right brand.

"Joe Boccio": Danny was talking about Joe Boccio, a "hard luck guy." Poor Joe was attacked on the Union Road job by a truck driver with a wrench, who sent him to the hospital for a long time with severe head injuries. On the downtown Buffalo road job, Joe got caught in the cab of a backhoe which caught on fire, and he just barely got out without being burned alive. On a job we did at the Buffalo airport, Joe was sent a cherry picker (small crane) that had a busted transmission before it came on the job. Joe was blamed for it and was laid off for a week. Danny recalled that Joe went to Chicago during the blizzard of 1978 to help with the snow removal. Danny remembered thinking, before they sent Joe, that they had better not send him because there might be another Chicago fire and Joe would get blamed for it. Or, there might be another St. Valentine's massacre and Joe would be one of those massacred. About two months after that conversation, Joe died of a heart attack.

"The Good Old Days": Danny and Sal were talking about the "good old days" when the men used to have to shape up for jobs. The contractors would come around and pick out the men they wanted. They talked about the old union meetings when an apprentice would have to ask for permission to speak. They had lots of apprentices then. Now, the only apprentices are the sons of members, and if they speak up at the union meeting, "the father gives them a cuff (smack with a cupped hand)." Sal said that in the good old days, the men would vote by lifting their shovels if they wanted to vote aye. Now, Danny chortled, the union has the meeting stuffed with hard guys who give their neighbors a jab in the ribs; the man responds by saying "A-i-e-e-e-!" in pain, up jumps his hand, and the chairman says it's an aye vote.

"Everything That Can Be Broken, Has Been": Danny was talking about a guy who was hit by a pile driver and had almost every bone in his body broken. Danny said he had so many plates in him that when it rained he "rusted." This guy was flagging on a road job when a woman ran two barricades and hit him. He was still waving his flag as he went down. When Danny bent over him, he said, "Don't worry, Dan, everything that could be broken, has been."

"I've Always Wanted To Do That": Danny told the story of the time he was working on a concrete sidewalk gang. They had just finished a good stretch of work—about 50 feet—and the concrete had just been broom finished and all the joints struck. Along comes this very old man. He stops at the edge of the newly finished sidewalk, looks from side to side, then looks at the men and with lips pursed and face determined he walks right through the entire length of the concrete sidewalk, leaving his footprints in the fresh concrete. Then, he turns to the men and says, "All my life I've wanted to do that and I never thought I'd get the chance. Now, I've done it and it feels good." And he walked off with a broad grin on his face. Danny said he and the other men could not feel anger at what the man did.

Tony again: Some inspector tried to talk to Tony. Tony said, "See my foreman." The inspector asked, "Who's your foreman." Tony said, "I don't know."

"Dirty Tony" (not Tony Z.): The men were talking about "Dirty Tony" who

was even dirtier than "Dirty Johnny." Danny said this man, who was a laborer, always smoked a cigarette which he never took out of his mouth. In the winter, the mucus would run out of his nose and merge with the cigarette. One time, Danny said, he was down in a manhole on a sewer line with Tony. It was time to go to lunch. The water was filled with fecal matter. When lunch time came, Tony threw away his cigarette, rinsed his hands in the filthy water, climbed out of the manhole, and sat down to eat lunch. He offered Danny half of his sandwich but Danny declined.

In the middle of the conversation, Tony Z. announced, "You're all lucky, I have to go to the dentist next week and you may not see me for half a day."

"My Sock Slipped": Tony said he knew it was quitting time at 4:30 because his sock slipped down to his toes and it was hurting.

"That Was No Bum": Frenchie said, "I dropped a piece of pipe and some bum picked it up for scrap." Tony replied "That was no bum, that was Johnny P."

"What's a Shovel Worth?":

Tony: "Sal confiscated my shovel."

Rickie: "You're being charged for it."

Tony: "All right, I found a pick. Can I get credit for a pick?"

Rickie: "O.K., a pick's worth half a shovel."

Tony: "Do you see pointed ears on me?"

"Killed By a Flying Shit House": The men were talking about the Niagara Falls power project where 25 men were killed. Danny mentioned one man who was killed by a portable toilet, called a "Johnny-on-the-spot." It was being lowered on to the job site when it slipped and landed on the man. Danny commented, "What do you tell his wife? Something romantic, like, 'Your husband was killed by a flying shit house.' "

"Angelo": Danny told about a laborer, Angelo, who was always in the shit house. Danny said Angelo sat on the toilet so much he had a ring around his ass. Danny said Angelo could somehow sense when Arthur, the big boss, was coming on the job. All of the men would be working when suddenly Angelo would start screaming at the other laborers to work faster or he would bawl out the concrete truck driver. Then, sure enough, at that moment, Artie would appear, be impressed with Angelo's performance, pat him on the back and say, "Keep up the good work, Angelo," and leave. Danny quipped that Artie must have had a special odor that Angelo could pick up.

"Poison Ivy": Danny mentioned that he always picked up poison ivy. All he had to do was "get near the stuff and it would jump up and grab me." Danny said he was working on a job, doing surveying in a field full of poison ivy. He had protected himself with gloves and long pants and felt protected. It was a hot day so he took off his shirt. He was still protecting his hands which were down in the bushes. Suddenly, a large black insect landed on his chest. Danny slammed his gloved hand, opened out flat, against his chest and yelled, "Gottcha!" His gloves, filled with the poison ivy, left their imprint on his chest in the form that the infection broke out. After that, he took a ribbing from the rest of the men, who kept slamming their palms against their chest, shouting "Gottcha!"

"Joe Cicero": The men remembered Joe Cicero. A truck ran over him. It was a tandem truck; one wheel caught one leg, the other wheel caught the other leg, and the two wheels pulled his body apart. Joey, the surveyor, dove under the truck and tried with his hands to hold Joe's blood vessels and organs together so they would stop bleeding. It was a futile effort, and Joe died on the way to the hospital.

"We Always Look Backward": Chuckie asked, "How come we do so many things backward?" Sal shot back, "That way we can see where we came from. We never look ahead, we might get confused."

"Labor Trainee": Tony commented about a labor trainee who was laid off. He said the man stood around so much like a statue he was covered with pigeon shit when he left. Tony said he had to ask the man to lift his feet so he could grade under him. Tony said, "When I asked him to hold a grade string for me, the man said, 'What, again?' "

"I Have To Start Stealing": Danny said he checked over his finances and came to the conclusion that the only way he could make ends meet was to start stealing. He said at the rate his wife was spending, it would have to be grand larceny.

"A Big Kiss": During the conversation, Bird, one of the gradall operators, came in to say so long, since he had been laid off. Danny jumped up and gave him a big, smacking kiss on the cheek and said, "I told you when you left I'd give you a kiss good-bye." Sal yelled to Bird, "You'd better get a tetanus shot!"

"Keep Your Shovel": After Bird left, Danny told the rest of the men in the room, "We'll be laying off the next two weeks. Any man who still has his shovel can stay."

"I'm Going To Finish the Job": Mike told about Johnny P. Johnny was shot up pretty badly in Vietnam and had a bad leg. He was also a fearless person, one of the few men willing to crawl inside a 24-inch pipe for hundreds of feet to check for leaks. Well, Johnny was on the Military Road job, in a 24-inch pipe, checking for leaks, when he suddenly got a cramp in his leg. He got panicky, thinking he might not be able to crawl out, and he scrambled back out of the pipe. When Artie, the boss, found out, he told Johnny he was not to go into the pipe anymore. Johnny got angry, started stamping his feet on the ground, and shouted at Artie, "I don't give a fuck if you fire me, Artie. I'm going back in that pipe to finish the job."

OCCUPATIONAL PRESTIGE

Job satisfaction comes, in part, from one's belief that his job carries prestige and is important to the community. Construction workers can point to the physical evidence of their work. In their day-to-day conversations, they talk with pride about significant projects they worked on—a power plant, an important thruway, a convention center, or a bridge. The larger the project, the greater the pride. One of the projects we worked on was featured in the trade magazine, *Architectural Forum.* I showed a copy to the men, and as they were looking at it, they described with great pride how they executed some of the work.

Construction men like to talk about work feats. I've listened to men talk about how many bricks they have laid in a day, or how many thousands of feet of road placed or thousands of cubic yards of concrete poured. Some tradesmen become famous because they are fast workers or capable of producing large amounts of work. One day I listened to the men talk about a man who was phenomenal at building manholes. The men said that even when the mason became old he could still wear out three laborers trying to keep up with him. One man, a superintendent, said he was stuck one day. He had to get a manhole built, so he went to see the old mason. The mason told the super to get all the materials at the excavation, to get him a laborer, and to wait at the hole at 5:00 P.M., after work. The super said he was waiting at the hole at 5:00 when, to his amazement, a bus pulled up in front of the hole. The bus door flew open and out stepped the old mason, dragging his tools in a bag. He finished the manhole in two hours, with the super and the laborer feeding him the block and mortar.

Construction workers also talk about the arduous and dangerous nature of their work with pride. Work that is dangerous carries prestige because the men who perform it are viewed as special. High steel workers are treated this way. As discussed previously, construction work has a high incidence of accident and death. Part of the culture of construction is the satisfaction of doing "manly" work, winning out over the elements and showing persistence in the face of adversity. Construction work is often hard and dirty, requiring one to work in foul weather (freezing or boiling), breathing dust, and being soaked to the skin by rain or snow. A construction man's hands are usually swollen and scarred, with a high incidence of broken or missing fingers. Construction workers enjoy the challenge of difficult tasks and the satisfaction that comes from doing a difficult job well.

JOB INSECURITY AS A SOURCE OF DISSATISFACTION

Unemployment is a major source of worry and dissatisfaction among construction workers. It is a cause of tension. During the winter of 1977–1978, on the downtown road project, we had one severe blizzard every two weeks. This caused frequent layoffs and many complaints from the men. Usually, management tries to keep its key people working, even in the winter on days when the weather is too bad to work in. Thus, it was a great surprise to us when we were directed to lay off some of our important people, men who had been with the company 15 or 20 years. One of these men, when he was laid off, remarked with bitterness,

> I come over here from the building division where I never lost a day and now I get laid off. I've worked through my lunch hour; I'm always on the job an hour early. I'm always pushing to make costs and this is the thanks I get. Fuck the roads division. I'm going to talk to A________ (the boss) and get back into buildings.

Unemployment compensation is based on a four-day week. Therefore, if a man works one day and is laid off, he makes $80 plus $120 for unemployment compensation. If he works two days, he loses his unemployment compensation and only earns $160. For this reason, many of the men ask to get laid off after working a

Monday or a Friday. Most construction unions now have supplements to unemployment compensation. Men with 20–26 weeks of work for the year are eligible, and benefits range from $25 to $65 per week.

One of the carpenters returned to the downtown road project after being laid off for one month. One of the men kidded him about his enjoying the time off, and the carpenter became angry. When he was asked why he was so high strung, the man replied,

> Because I've been off for a month. Christ, I was ready to go to work in a supermarket. Everybody tells me I'm a damn good carpenter but that don't mean shit. This shit's been going on for years. If things don't change I'm going to open a little bar. I don't care how little I make. At least it'll be steady.

Unemployment is higher in construction than in any other industry in the United States. In August 1977, general unemployment in the nation was approximately 7 percent, while in construction it was 12 percent (*Construction Equipment*, August 1977:11). Unemployment in construction in February is usually at least twice as high as unemployment in the nation. The high hourly wage of the construction man is partially offset by the high rate of unemployment in the industry. John T. Joyce, National Secretary of the Bricklayers Union, commented on this aspect of construction work (Joyce 1973:10):

> Just how affluent is the construction worker? Hourly earnings do not give an accurate picture. . . . According to Labor Department studies, the average full-time construction worker works only about 1,500 hours at his trade each year. . . . A major cause of underemployment among building tradesmen, is of course, the tendency to close down construction jobs in bad weather. But other liabilities of this kind of work not only curb a construction worker's earnings but also cause him many other problems. Since construction workers are only employed for a given job or project, they have frequent periods of unemployment between jobs. Unemployment rates in construction are historically double the national figures—even when the economy is strong.

Mike Cherry, an ironworker, tells what weather can do to a construction man's income (1974:78):

> In 1970 I worked in New York City from March 1st on. That was a famous year among construction people, because the winter was virtually snowless and the spring was almost without rain. In those ten months I lost a day and a half to the weather. The following year I lost forty-three days, which comes to about $3400. Thus, in spite of the fact that I worked two months more, I earned about the same.

Unemployment in construction has been a source of concern to the industry as well as to the government. A number of proposals have been offered to control the fluctuations of activity so characteristic of the industry. One suggestion has been to subsidize firms which accelerate rather than reduce their building programs during the winter months (HUD 1973). This is the practice in Norway and Canada. Another proposal is the scheduling of contract awards for the spring of the year, so that construction projects can get a good start and maintain high levels of activity and employment (U.S. Bureau of Labor Statistics 1967). There have been suggestions to help construction workers find other jobs during the winter months. Another idea is to use operating engineers to run snow-removal equipment in the winter (U.S. Bureau of Labor Statistics 1970). A recently passed act contains a provision that will steer government construction contracts to officially designated labor surplus areas—those with unemployment rates 30 percent above the national average (*Engineering News-Record*, November 7, 1977:3). Some of the suggestions to combat construction unemployment involve technological changes to permit work to continue through winter. Some of the ideas are use of rust-resistant steel that does not require paint; additives in concrete so it can be poured in below-freezing weather; and special methods to heat concrete or masonry.

Job insecurity is a major source of job dissatisfaction among construction workers and should be viewed in combination with all the other factors that make for high job satisfaction in construction. While it is a source of worry and concern, I do not believe it overshadows all the other positive factors which foster satisfaction on the job for construction men.

In conclusion, it might be instructive to look at what is probably the most comprehensive study on job satisfaction. This is the University of Michigan Survey Research Center study, "Survey of Working Conditions (November 1970)" (1971). This study contains the best comparative data assembled to date on job satisfaction. It has comparisons by age, occupation, personal income, industry, race, occupational group, education, and sex. In the study, "contract construction" had the lowest percentage of "workers expressing negative attitudes toward work"—only 5 percent. The data, by industry, are as follows (1971:408):

Industry	*Number in Sample*	*Percentage Dissatisfied*
Contract Construction	123	5%
Manufacturing	381	17%
Wholesale, Retail	274	23%
Services	397	12%

8/Overview of the construction industry

We have examined the behavioral patterns and cultural views of construction workers. It will now be instructive to examine these factors within the context of the construction industry, which will show that construction worker behavior is appropriate for the social organization and technology of the industry.

IMPORTANCE OF THE CONSTRUCTION INDUSTRY

The construction industry employs more than 4 million workers (Census of Construction Industries, 1972) and accounts, directly and indirectly, for 13 to 14 percent of the United States Gross National Product (U.S. Department of Commerce Construction Review 1976).

The construction industry is of major importance to the functioning of the United States economy. Variations in the level and rate of investment are considered by economists to be significant in determining total economic activity. In the long run, the volume of investment in buildings, roads, dams, and other structures largely determines the capacity of our national economy to produce and distribute the goods and services needed by consumers. Mattila and Gaumitz stated (1955:6):

> An understanding of the fluctuations of investment activity in this industry (construction) is necessary for an understanding of the fluctuations in income and employment in the total economy.

Here are some additional facts pointing to the importance of construction in the United States. Construction employs 15 percent of the total skilled labor force in the United States (Mills 1972:4). Construction accounts for 66 to 80 percent of the output of the lumber, stone, structural metal, clay, plumbing, and heating industries in the country, and it accounts for one-sixth to one-half of 12 additional industries (Bureau of Labor Statistics 1970:1642).

FIVE MAIN CHARACTERISTICS OF THE CONSTRUCTION INDUSTRY

Summarized below is an outline of the five main technological and organizational characteristics of the construction industry. These characteristics provide the basis for the culture and behavioral patterns that are "appropriate" on construction projects.

Uncertainty Risk and uncertainty are major factors in the construction process. It is so widespread that the American Society of Civil Engineering held a week-long conference in January 1979 on the subject of "Construction Risk" (*Civil Engineering* 1979). The very concept of a construction project is beset with uncertainty. Will the owner go ahead? Will he receive his financing? Will the labor and materials needed be available during the entire course of the project? Will the project receive zoning board approval? Will the project be scuttled by a legislature or a citizens' lobby or by an anthropologist who discovers that it threatens an archeological site?

It takes so long to build most major projects that something unexpected is bound to happen, and it usually does—a strike, a material shortage, an economic recession, inflation, cutoff of funds, a rash of bad weather, a labor shortage—even a revolution, as many construction firms in Iran found out.

Even if the project gets through all the preliminary uncertainty, there are still the design process, the bidding process, the awarding of contract, and the whole charade of trying to set a project completion time. Then you start production, not knowing what is in store for you. Conditions may be quite different from what you saw on the plan. Even after you bid a job and think that you're the successful bidder, you may not get the job. The owners or authorities may throw out all bids for a number of reasons—the bids came in over the engineer's estimate; or the engineers and owners want some other contractor and will find some excuse to throw out the bids. This happened to us in an out-of-town bid. The owners took the bids three times before they got the contractor they wanted. The rest of the bidders knew what was going on and withdrew from the bidding.

In addition to everything discussed above, there are three features in the construction industry which contribute to uncertainty:

1. Economic cycle variations.
2. Seasonal uncertainty.
3. Potential causes for setbacks.

Every construction worker knows that the job he is on has a limited time period, on the average two or three years. He is also aware that his own craft or trade may not be needed continuously. He may be called to the job, stay for a limited time to perform his work, and then be asked to leave and return later when his craft is needed again. He may or may not then proceed to another job. Flexibility, mobility, and constantly changing job assignments are all part of the way of life for most construction workers.

On any given day when he comes to work, a construction man may or may not work. Even if he starts working, he may be rained out during the course of the working day. There is also the unpredictability of work assignments. If a machine breaks down he may find himself in another crew, at another location, doing a different task. Often, a construction worker does not know from day to day what his assignment will be. If he is a bricklayer, he may be at ground level one day and ten stories high on a scaffold the next.

Thus, there are no repetitious tasks in construction. Flexibility and changeability are the norms. The men come to expect it. They accept it because they deal face to face with their foremen and superintendents, and changes are explained to them. In addition, they see how the overall project is progressing and they understand how their work fits into the total pattern of a project. They have a strong notion of where they and their skills fit, so the constant movement and change are things they can understand and react to.

Hand Tool Technology Construction is one of the few industries in the United States which still relies heavily on hand tools and a handicraft technology (Haber and Levinson 1956:19). Where else in a major industry do workers own their own tools? Most craftsmen in the industry are required to provide their own hand tools. They do not own the heavy construction equipment, material-handling equipment, or power tools. But whatever hand tools are needed they provide themselves. This is what gives the craftsman control over the work process.

The handicraft technology relates to the craft organization of labor in the building industry, whereby about 50 skills are organized into 20 craft unions. A mason from ancient Egypt or from medieval Europe would be quite at home on today's building projects. He would be impressed by some of the material-handling equipment, but the actual laying up of masonry units would be quite familiar to him. The same would be true of the ancient carpenter.

Anyone on or near a construction project will quickly notice the high percentage of pickup trucks owned by the workers. Many of these trucks have tool boxes custom-built into them, or have racks to carry pipe and lumber. Some have small compressors on them or an oxyacetylene burning and welding outfit. The trucks and tools reflect two things about construction workers—they own their means of production and they can easily become their own bosses.

Ownership of tools and skills by the craftsman leads construction management to rely on the skilled worker to create the product. The skilled craftsman asserts that he knows better than anyone how to perform his work. For its part, management expects the craftsman to perform his job with a minimum of supervision. Management's role is often no more than telling the tradesman what, where, and when to produce. Even the engineer and architect must rely on the craftsman to interpret their plans and carry out their intent. This is in sharp contrast with factory work, where the industrial engineer aims "to tell every worker in the corporation exactly how to do his job down to the last detail" (Walker 1962:136).

The Informal System Although construction is increasingly becoming formalized, with a stress on documenting decisions and lots of letter writing, the predominant mode of operations on the job site is still informality. Face-to-face relationships and acceptance of verbal decisions without formal procedures are normal methods of operating a construction project. The formal system assumes that everything is clear on the plans and in the specifications and contract. This is never the case. Projects would never get done on time if every change had to be channeled through formal procedures; if mistakes were not caught in the field and on-the-spot solutions were not sought; if thousands of compromises were not worked out every day.

In construction, you need a crisis-oriented personality, particularly if you are directing work in the field. Things happen which require quick decisions: equipment breaks down and men have to be shifted; a wall may start to shift or give way and has to be quickly shored up; a trench in which you're installing pipe collapses; the concrete plant or asphalt plant which was supposed to deliver your material broke down; someone got hurt; the steel is the wrong size; "where the hell are the windows?"; a torrential downpour wipes out your site work or your roofing work for the day; the temperature drops below freezing and you can't pour the bridge deck concrete, knocking your schedule to hell; a key subcontractor who was supposed to show up, didn't. All these things require shifts in plans and decisions and you don't have time to follow the formal lines of communication—you have to act on your own, and you're expected to. Sometimes, you have to say, "It's no use working today," and you send all the men home—a hard decision. The men lose a day's pay and you fall behind in your schedule.

Construction work is, by and large, an informal, flexible, operate-by-the-seat-of-your-pants operation. Of course, you can draw up an impressive bar chart or flow diagram showing "just how" the work is supposed to progress. These charts are helpful in organizing your thinking. But the deviations from them can be massive, and anyone who tries to stick rigidly to some preplanned chart will look ridiculous.

Still, the jobs get done and by some miracle, close to the completion date. But it takes a lot of turning and twisting and manipulation of men, materials, and equipment, and a lot of luck with the weather. Projects come close to completion dates because many people are willing to do a lot of bending and compromising. A formalized, regular, step-by-step process is out-of-step and completely unworkable on most construction projects.

The Subcontracting System The subcontracting system is suitable for the construction industry. It permits the kind of flexibility required whereby various mixes of contractors and crafts must be mobilized to suit the unique requirements of a particular project. Then, when the project is over, the contractors disperse. Subcontractors make possible the rapid mobilization and dismissal of crafts for a project as changes in the kind and volume of construction demand vary.

The general contractor cannot do everything. He doesn't have the know-how and he doesn't have the manpower or supervision. He must rely on the subcontractor. Actually, there are far more specialty subcontractors—twice as many—as there are general contractors. Fifty percent of construction contractors are specialty subcontractors; 25 percent are general contractors; 12 percent are heavy and highway contractors; and the remainder are developers (Census of Construction Industries 1972).

The general contractor shares administration and supervision with his subs.

This diffuses project administration and the network of responsibility is more horizontal than vertical. The general contractor deals with subcontractors who are independent and involved in other projects. The general must therefore request, and sometimes threaten, a subcontractor to get him to perform. But he can't fire a sub as he can his own employees. The subcontractor has a contract and can sue the general contractor if the latter breaks the contract arbitrarily.

Subcontracting fosters a sense of democracy on the job. Every trade maintains its individual importance and standing. Trades must respect each other since they depend on others before they can do their own job. Even if a contractor is small, he still receives the respect of his associates because his trade is essential. On the downtown Buffalo road project, our sidewalk subcontractor was so small and weak financially, that we, the general contractor, had to buy all his materials and put his men on our payroll. The boss worked along with the men in his crew. All the other trades, as well as the State Department of Transportation, had the highest respect and admiration for this contractor because of the excellent quality of his work.

Subcontracting also leads to family-like relationships between the general contractor and his "family" of subcontractors. Many general contractors use the same group of subcontracting firms over the years. General contractors like to repeat with good subs. If the relationship is continuous over a protracted period, the general contractor and his subs will often develop a close personal relationship. They will socialize and engage in recreational activities together. Similarly, the workers for each firm will come to know each other and will work well in the field. Many of them will become close friends, just as their respective employers have. It makes for a trusting and responsible relationship on the project. Good friends do not let each other down, in their private or their work lives.

The Localized Nature of the Construction Industry Every locality in the United States has its own construction industry. The reason is the nature of the construction product. It is bulky, costly, and unique to a particular site. Structures are designed to fit a particular piece of land. No factory can produce office buildings, power plants, schools, dams, or bridges and then ship the finished project to a site.

The 20 largest firms in the construction industry do not constitute more than five percent of the receipts of the construction industry. The reason is that construction is largely an industry of small, local contractors. Construction has more than 900,000 firms, of which half have no employees—that is, they are run by working bosses (Census of Construction Industries 1972). Eighty-eight percent of receipts from contracting originate in contractors' own home state (Census of Construction Industries 1972). Most of these are also from the home town or city.

Since construction is an industry of small, local firms, almost everyone in the local industry knows each other. This includes the labor force, the contractors and subcontractors, the local supply houses and equipment rental companies, the architects, and the engineers. This group of firms and workers, of suppliers and renters and professionals takes on the characteristics of a community. It tries, with much success, to keep outsiders out of the area, since there is a limited amount of construction work in every locality and the local community naturally wishes to maintain it for themselves. It is in the interest of all parties to do so—the labor unions, contractors, suppliers, banks, local architects and engineers, supply houses, and the construction workers.

THE PRICE OF STRUCTURES

Construction is unique, differing significantly from manufacturing, service, and transportation. Unlike manufacturing, where the producer decides what to make, construction products are determined by the customer. Pricing in construction is also unique. Unlike many industries, which issue a catalogue of their products with accompanying prices, there is no catalogue of prices of buildings or structures. In construction, it is only when the nature of the desired product is determined through designed plans and specifications that it can be priced. Each building structure is

Figure 14. One can see from this picture why construction structures can be described as bulky, costly, and unique, custom-built to fit on a particular site in a particular area. These are the carbon storage tanks from the North Tonawanda sewage treatment plant project. Cost: $24 million. The tanks were 50 feet in height, and eight of them were enclosed inside the carbon filter building. Earl Kostuk, to whom the book is dedicated, designed the forms for these tanks. (Photo courtesy of John Kaiser, Roy Crogan & Son.)

priced separately and distinctly in the form of a bid for that particular project. Bidding arrangements can vary, depending on whether the contract is let on a lump-sum, cost-plus, upset figure, or construction management basis. Many projects have select bidders lists prepared by the owner's architect or the owner. Contracts let on a competitive basis have provisions for extras and credits and other contingencies that might arise during the construction period.

FOUR STAGES OF THE CONSTRUCTION PROCESS

The construction process passes through various stages. Assuming that all of the preplanning and financing arrangements are accounted for, there are four main stages in construction:

1. First there is the client, public or private, who decides he wants or needs a particular structure built. In rare cases, the client acts as his own designer and contractor.

2. Next, the client retains an engineer or architect to design and estimate the cost of construction. The engineer advises the client on what the structure must contain to meet the client's needs. If the project is particularly complex or specialized, the architect will use his own consulting experts. These, in turn, will help design those aspects pertaining to their specialty. The engineer, architect, or specialty consultant will usually oversee the construction to make sure it is in accordance with his intended design.

3. Once a proposed project is set forth in plans and specifications, it is advertised for competitive bids. Some contracts are negotiated, but this is not common. Competitive bidding usually takes 30 to 60 days. In some cases, such as sewer and utility work or road building, general contractors can undertake the entire job themselves and formulate bids on their own. Most other construction does not conform to this pattern. Some general contractors might account for 30 to 40 percent of the work if they do their own carpentry, concrete work, masonry, and structural steel. General contractors must secure bids from other trades and suppliers before they can submit their own bids. Some specialty subcontractors subcontract some of their work, and they too must take bids from others. For example, a heating, ventilating, and air conditioning subcontractor must take bids from equipment suppliers, sheet metal subcontractors, temperature control subcontractors, and insulation subcontractors. All of this solicitation of sub-bids delays the bidding process. At every level, careful estimates must be made before the final figure is submitted. Bidding costs are high and must be included in the annual costs of doing business in the construction field.

Competitive bidding can be disadvantageous to the industry. The downward pressures of competition may lead a contractor to substitute an inferior product for that specified. He may try to avoid union labor. A contractor with a low bid may turn to fly-by-night subcontractors who are willing to work at any price but are unqualified to perform the work. This leads to subcontractor failures. Bid bonds keep out unqualified bidders, but it does not prevent a successful general contractor from shopping for unrealistically low prices. Contractors try to protect against these practices by forming associations, but many firms do not belong to them. Fly-by-night contractors spring up constantly, since it takes little capital and it is relatively easy to enter the construction industry. Ultimately, it is the construction workers, through their trade unions, who have established some uniformity through the imposition of uniform labor rates and conditions for a particular trade in a particular locality in the country (Slichter, Healy, and Livernash 1960).

4. The fourth step in the building process is to build the structure for the price of the contract. The engineer or architect will designate a representative to oversee the project in the field to make sure work is performed in accordance with plans and specifications. Monthly payments are made to the contractor to permit him to pay his subcontractors and suppliers and to meet his own monthly costs. At the conclusion of the project, the engineer or architect prepares a punch list of incomplete or inadequate items of work which the general contractor and his subcontractors must rectify. The owner withholds a certain percentage from each payment to guarantee that the contractor will have an incentive to complete all work properly. When the structure and all of its systems are accepted by the owner, a date of occupancy for the structure is established and final payment is made to the general contractor. Most contracts have a one-year guarantee which requires the contractor to repair or correct any defects during that period. At the end of the first year of occupancy, if all corrections have been made, all contractual obligations are deemed to have been met. The only exception is that some manufacturers of equipment give longer warrantees on some of their components, like a boiler or an air conditioning compressor.

THE LABOR FORCE

Eighty percent of the labor force in construction are blue-collar workers. The building construction labor force is composed of more than 20 crafts and many more specialties. Many contractors directly hire only one or two of those crafts. The number of workers and crafts hired depend on the type of work and the geographic area in which the contractor operates. Finding a job in construction is a relatively simple matter when construction activity is high. Some projects usually are starting as others are finishing, and some contractors are hiring as others are laying off. The amount of time off for the worker between jobs may be large or small, depending upon the amount of construction in the area. Seasonal unemployment, however, is ever-present in certain trades, even in years of high construction activity. As the rate of activity declines between November and March, workers are being hired for new projects at a slower rate than other workers are being laid off from projects approaching completion. From late fall until early spring, lost time between layoff and new jobs may be considerable even when activity in an area is high.

Workers are ordinarily hired by a foreman, who selects applicants either at the job site or by contacting the office of union locals who represent the needed crafts.

The worker is subject to being laid off at any time, either permanently (as the work for which he was hired approaches completion) or temporarily, with instructions to return at a stated time when his services will again be needed on the project. Each trade group comes to the job, performs its portion of the work, and leaves, as the work peaks and decreases at varying rates. Trades leave and return at differing rates. The production process involves constant adjustments and jockeying of the labor force to mesh one trade with another.

Many construction workers do not expect to work permanently for one contractor. As compared with industrial workers, almost twice as many construction workers will be likely to work for more than one employer in the course of the year (Bureau of Labor Statistics 1970: Bulletin 1642:6). They are more attached to their trade than to any employer. Knowing the nature of their industry, they take it for granted that they must be prepared to move, not only from site to site but from city to city. A construction worker may migrate to an area with better long-term employment prospects. For a union member, such a move involves transfer of membership. Like other unions, building trades unions are organized through chapters known as locals, each having jurisdiction over a designated geographical area. A member moving elsewhere usually can exchange his membership card for a card in a local at his new location. Thereafter, he is a member in the new local. Should he return to his previous place of work, he must obtain another transfer.

A construction worker usually leaves an area as a matter of choice, although there are exceptions. When a construction project is undertaken in a distant locality, comparatively few of the needed workers may live within reasonable commuting distance. When general construction activity is high, employers have difficulty manning isolated construction projects. When general construction activity is poor, men with family responsibilities must choose between the prospect of unemployment at home or a relatively steady job in a distant or isolated community.

The construction labor force is characterized by a high degree of trade specialization. This creates the need for a substantial degree of interdependence between crafts, requiring a constant shifting of the labor force on and off the projects. Employers greatly value the ability to hire and lay off men as job conditions demand. Equally important is the ability to hire men at a predetermined wage scale, so the profit on a job is not eaten up by wage changes after a bid price is established. The flexibility that the contractor requires translates into job insecurity for the construction worker. Therefore, he turns to his union to limit the effects of job insecurity that stem from the nature of the labor force and construction job conditions.

ROLE OF THE CONSTRUCTION UNIONS

The peculiar economic and employment conditions in construction place employers and unions in a much more intimate relationship than that found in many other industries. Collective bargaining agreements are areawide; they cover the contractor and the worker whenever they work or operate in the area. This guarantees a continuing relationship between union and employer, more so than between worker and employer.

As noted, flexibility is of critical importance to the contractor's need to expand or contract rapidly as market or project conditions change. The craft union supports this specialization and also performs functions that stabilize the industry. It enforces standards of work and compensation, participates in apprentice training for the crafts, and refers workers to contractors upon the latter's request. At the same time, it allows the direct employment relationship between contractor and construction worker to remain casual. Thus, the craft union allows flexible employment relationships to exist for the skilled work force while maintaining stability through its enforcement of uniform wage rates.

In an unstable industry, construction unions function to provide a stable source of skilled craft labor for contractors while preserving job opportunities for workers. One of the mechanisms through which this is achieved is the union hiring hall. Craft unions provide a placement service through their hiring halls that benefit both contractors and workers. When a man seeks work, he visits former employers at job sites and contacts superintendents or foremen he knows. If this does not lead to a job, he resorts to the union hiring hall. The hiring hall serves as a central pool of labor for each craft or trade. The craft union acts through the hall as a clearinghouse for craft workers in their jurisdiction. To secure their central place in the industry, the unions strive for a monopoly position. They try to negotiate a closed shop and an exclusive hiring hall arrangement. This has met with surprisingly little resistance from contractors.

As stated previously, construction operations involve the movement of firms and the hiring of qualified workers from project to project as new work is undertaken. The flow of any firm's work may not be continuous, since the securing of new jobs may involve periods of inactivity. Once a contract is awarded, the general contractor hires labor for that particular project as it is required in the successive stages of construction. The subcontractors involved must plan their work to mesh with other trades and the overall schedule of the general contractor. When each stage is complete, the labor force leaves for another job or is laid off. The general contractor seldom maintains more than a small permanent crew of key men. They depend on the unions to supply the skilled craftsmen as they are needed on short notice. The union functions to collect employment information, dispatches men to projects, and distributes work when it is scarce. A contractor finds that hiring from the union hall is a distinct advantage. In most cases, the skills and experience of the men sent by the union will meet the necessary standards. In effect, the union certifies their qualifications. They attempt to match the requirements of the contractor with the available supply of labor in the hall. There is no requirement to retain a particular worker if he proves unsuitable or unqualified.

There are also risks for employers who use a closed-shop, hiring-hall setup. The unions may not, in fact, have qualified men in the hall. They may not give the contractor an opportunity to ask for men he has employed to his previous satisfaction. Unions often employ a system of rotation, which serves to protect their less qualified members at the expense of the qualified. They may use a system of favorites. Workers close to or related to the incumbent administration may be favored. Contractors who give the union trouble may get all the "butchers." The union may try to strengthen its bargaining position by limiting entry of new members into their trades. For all of the above reasons, many contractors, while agreeing to the closed shop, will insist on the right to do their own hiring or use the union's hiring hall as they see fit.

Whether contractors accept the union hiring hall or not, the closed-shop principle is of benefit to contractors and workers. By placing a wage floor under the competitive forces at work in the industry, it prevents competition from getting out of hand. This lends an element of stability to the industry which might otherwise be lacking. A uniform wage rate throughout a local market area is advantageous. Each contractor can submit only one bid and is presumably ignorant of the bid prices submitted by others. A major factor in the costs of each competing contractor is the total amount of wages paid for each trade. If all contractors know that their competitors must pay an identical wage rate, this removes a large item from the sphere of competition. In the absence of unions, there is little reason to believe a single wage rate for a particular skill would prevail. Contractor competition based on paying differential wage rates would become intense, and the resulting instability would work to the detriment of all involved in the construction industry.

Not all crafts, sectors, or geographic areas are unionized. But all contractors, union or nonunion, are influenced by the prevailing labor policies of the unionized sector. Wages in the union sector influence what nonunion contractors must pay. Union organization in the mid-1960s accounted for 68 percent of all construction workers employed by contractors and subcontractors (Bureau of Labor Statistics 1970:Bulletin 1656). Wages tend to be higher in the unionized sector, sometimes by a considerable amount. These wage differentials not only reflect the strength and bargaining power of the construction unions, but also the skill levels and the nonresidential market where unionization is highest.

9/Summary and conclusion

INTRODUCTION

I have presented the idea that the social organization of the construction industry, which is the result of the technology and the characteristics of the industry, lead to construction worker behavior which stress the following:

1. Autonomy and independence of the craftsmen.
2. Control over the work process by the workmen.
3. Decentralized decision making on the construction site.
4. Loose supervision of craftsmen and tradesmen.
5. Integrated work groups.
6. A large degree of work satisfaction.

THEORETICAL ORIENTATION

The theoretical orientation of this book is based on the idea that construction workers develop a distinctive culture based on their work environment and the ideas and values they derive from performing construction work. Otterbein has defined culture as "everything that a group of people thinks, and says, and does and makes" (1972:1). The people in this book form an occupational group—construction workers. I have not dealt with "everything" they think, say, do, and make, but only when they do these things in their work environment—the construction project. Based on the social structure they share in their work environment, construction workers constitute a distinctive subculture. They have a set of ideas, values, beliefs, and behavior patterns that stem from their work, which they learn and transmit to new members of their group.

Another theoretical orientation of this book is the notion that occupation is an important factor in modern society in the formation of beliefs and attitudes. An extension of this idea is the concept that occupational groups also develop distinctive and identifiable subcultures based on their work and work places.

THE SOCIAL AND TECHNOLOGICAL BASE OF THE CONSTRUCTION INDUSTRY

The technological base of the construction industry is mainly hand labor and handicraft technology. For this reason, there is great reliance on the individual workman. Labor, rather than machinery, is the key to production and output in the construction field. Construction is thus a labor-intensive industry. Its growth means the growth of employment.

The social base of the construction industry is decentralized and locally oriented. This tends to limit the size of firms and with it, the growth of large bureaucratic organizations. It also tends to stress informality, personal relations, and community-like networks through the boundaries and limits set by local and decentralized organizations.

The construction industry is characterized by instability and wide fluctuations of activity. This is reflected in the culture of construction workers where job insecurity is a way of life. The large degree of flexibility and variability in the industry makes mass production unfeasible. This relates to the reliance on hand methods and craftsmen rather than automatic machinery to create the product. Reliance on hand methods and craftsmen engenders independence and self-respect among construction workers.

The social organization of the construction industry is flexible and mobile with constantly shifting workers. This reduces loyalty of employees to a particular employer. Ultimately, construction workers turn to their union and fellow workers for security and support.

Construction workers display a special loyalty to their craft. The craft organization is a subsystem in the social organization of construction and it fosters craft consciousness. At the next level is occupational consciousness that ties all construction workers together. At the higher level is blue-collar consciousness, in which construction workers identify with people who work with their hands or engage in manual trades. Ideologically, one finds little "class consciousness" among construction workers regarding worker–boss differentiation. The high wages and life styles of construction craftsmen is not conducive to hostility toward employers as a class or toward capitalists as a social group.

Construction is a complex industry, involving constantly changing environments and conditions. It requires flexibility in its labor force and contractors. Thus, the subcontracting system, which can adapt quickly to changes, is well suited to the social organization of the construction industry. Construction also relies heavily on worker skills and the human factor, which is also more adaptable to changes than investment in large, fixed capital. All of this is reflected in the culture of construction work which tends to be loose, informal, and personal.

Uncertainty and imprecision are important elements in the nature of the construction industry. It is a factor in how well or poorly a production task can be controlled. As the data presented in this book has shown, uncertainty is high in construction because of environmental factors (weather and climate), the variability of the construction site and the long period of production, and the many steps involved in the building process. This reduces control and reinforces the kind of loose, day-to-day, decentralized, on-the-spot kind of administration of construction work that takes place on the job site.

CONSTRUCTION WORKERS' BEHAVIOR

One of the key elements in construction work is the pattern of personal relations in the work process. The use of small crews, hand tools, and skilled hand labor fosters personal relationships in organizing the work. It gives the construction worker a measure of control over the work and a sense of his own importance and involvement in creating the product.

Controlling the work process is a key element in construction behavior and lays the basis for the large measure of job satisfaction that construction workers enjoy. Since the craftsman is permitted to work out the organization and methods of accomplishing his work tasks, he controls the pace of his work as well as the planning and solving of work assignments. This enhances the role and prestige of the tradesman, which may be raised or lowered by the behavior of individual journeymen, but remains basically high. Construction men behave on the job as independent units capable of carrying out their tasks with only general communications as to the work required, leaving all the intermediate steps to their own organizing ability and ingenuity.

The uncertainty factor, so high in construction, leads to reliance by the men on their union rather than their employer for security. Thus, construction workers behave toward their craft and their union with more loyalty than to their employer. This leads to behavior which is independent and assertive.

As part of the craft nature of the industry, the traditional work ethic is still strong in construction. The behavioral patterns of construction workers include getting to work on time, following instructions and orders, cooperating with other crafts, having pride in workmanship, being honest about one's work, caring for tools and equipment, and being willing to perform difficult or dangerous work. This is an ideal cultural norm never fully lived up to. But the fact that it is the norm affects the consciousness and behavior of construction workers who do, in the main, strive to live up to it.

Not only are work task activities conducted on a personal basis, but recruitment to the industry is also carried out through family and friendship relationships. The most frequent and surest way to get into construction is to know someone in the industry. As the data attest, father–son and brother combinations are still prevalent in the industry. The personal basis of recruitment reinforces and supports the personalized nature of the work process.

One of the cultural characteristics of construction workers is their love of outdoor sports and activities. They also enjoy kidding and horseplay, both on and off the job. It serves to create group cohesion and familiarity, which reinforces the personalized nature of their work relationships. The men know each other well enough to refer to each other's personal and private lives. Because they are close socially, they are willing and able to exchange insults in a friendly, kidding manner, similar in concept to what anthropologists know as the "joking relationship."

INDEPENDENCE AND AUTONOMY

Construction workers are not part of the type of bureaucratic administration that exists in most large industrial and business organizations in this country. This is due to the technological and social organization that characterizes the construction industry, which permits control of the work process to be held by the craft worker. This process results in a cultural pattern of independence and autonomy among construction craftsmen.

Ownership of the tools for production by workers is extremely rare in our society. Construction is one major industry where this situation exists. Union agreements in construction spell out which tools are to be provided by the craftsman and which furnished by the contractor. Most hand tools are owned by the tradesman and power tools are the property of the employer. Ownership of tools gives the construction craftsman the alternative of going into his own business. This increases his independence and autonomy, since the employer is not only dependent upon the craftsman's labor but also the craftsman's proprietorship of the tools necessary for production. The alienation of workers from their work because of their separation from tools and means of production is not felt by construction workers. Since they have the tools as well as the skills, they can put themselves to work independently of their employers if they so choose.

The existence of worker autonomy as a strong element in the culture of construction workers leads to conflict when management encroaches on this autonomy. Skilled craft workers believe they are the only ones capable of judging their own work and will resist attempts by management or owners to interfere with the process of self-evaluation. A competent journeyman knows his talents are sought after by contractors, and so he is not afraid to quit a job if he does not like the way things are run. This makes him feel independent even in the face of the insecurity in the construction field. The independence of construction workers is supported by the union, which controls the work rules, the training of apprentices, and entry into the industry. Since supervisory personnel belong to the union, contractor management finds it difficult to exercise autocratic power over workers through their supervisors. Management is ultimately dependent upon the craftsman's responsibility to his work, thus reinforcing the independence of the craftsmen.

Every trade in construction is considered important. No building can be completed without the contribution of every trade. This produces a kind of democracy on a construction project for each craft. This democracy extends the sense of independence and autonomy, since every tradesman knows his contribution counts for something significant in the total construction process.

The hiring and firing process in construction is reflective of the autonomy enjoyed by the men. Construction workers are hired directly for a project, sometimes on the project, without the involvement of management. Employment is based on personal contacts. Workers feel a sense of obligation to those who hire them. Since construction workers secure their jobs through personal contacts and union membership, they feel less tied to their employers and more to each other, thus giving them a collective sense of independence in relation to the contractor.

Autonomy and independence among construction workers are enhanced by the worker's right to refuse to work in inclement weather or unsafe conditions. Workers, not management, make the decision whether or not to work. This right is backed by the union agreement and, in the case of unsafe conditions, by federal regulations. This situation is in contrast to factory rules where refusal to work is usually a cause for dismissal. The fact that construction workers can refuse to work, with just cause, provides them with an additional sense of power and its consequent feeling of autonomy.

JOB SATISFACTION

The culminating idea of this book is that the main patterns of technology, organization, behavior, and culture in the construction industry lead to a large degree of job satisfaction for construction workers. The population studied manifested their satisfaction with their jobs through their conversations and by objective criteria such as the virtual nonexistence of absenteeism, lateness, or job grievances.

The measurement of job satisfaction has been the subject of a long and complicated research literature. No attempt was made in this book to use an experimental method to test job satisfaction among construction workers. Instead, a number of factors identified by researchers as correlating with job satisfaction were used to ascertain if they were present among construction workers.

One of the critical factors in job satisfaction is supervision. An examination of supervision among the construction men revealed most of the positive elements associated with job satisfaction. These were:

1. Supervisors who treated their subordinates as human beings rather than cogs in a production machine.
2. Supervisors who were democratic in style rather than authoritarian.
3. Supervisors who were employee-oriented rather than production-oriented.
4. Superintendents and foremen who considered themselves part of the work group rather than management.
5. Supervisors who belonged to the same unions as the workers and were often friends or relatives of the people they supervised.

Participation in decision making and control over the work process is important in providing job satisfaction. Various studies were cited to support this idea. Data in this study showed how crews operate autonomously, fulfilling their work assignments by making independent decisions. Autonomy in the work process, together with a nonbureaucratic administration, increases job satisfaction for construction men.

Studies among construction workers and others have shown that when men work in teams and interact with others on the job, their job satisfaction goes up. Conversely, job satisfaction is low when workers must work alone. Workers cite congenial peer relationships as among the important characteristics of a good job, and they often quit jobs that prevent contact with others. Construction workers are fortunate in that the technology of their work leads to the use of teams and crews to perform the work tasks. Human sociability is an intrinsic part of the work environment on construction projects, which function like minicommunities during the life of the building process. Integrated work groups bind construction men together in their work assignments. Their tasks require cooperation between trades and within crews. This atmosphere generates feelings of job satisfaction that flow from the knowledge that one is an accepted member of a social group.

An extension of the concept of integrated work groups is the occupational community. The subject of whether or not construction workers actually do constitute an occupational community needs more research. I have only shown that construction workers exhibit some of the characteristics of occupational communities—the mutual sharing of their work and nonwork lives; the development of self-images and value systems, in part, from their work environment; and the development of a sense of prestige from belonging to an identifiable social group.

Not all aspects of construction work are satisfying. A major source of dissatisfaction is the high rate of unemployment in the industry. It leads to insecurity and lowers the annual wages of those in the field. Another source of dissatisfaction is bad weather, which leads to temporary layoffs and is unpleasant to work in. Danger is not really a source of dissatisfaction. It is accepted as a part of the occupation, and construction workers get satisfaction from overcoming dangerous or hazardous work.

In spite of the sources of dissatisfaction, the elements that favor job satisfaction outweigh the negative ones. In general, construction workers enjoy a high degree of job satisfaction which is reflected in their attitudes and their behavior.

A FINAL COMMENT

This study has dealt with a population not usually researched by anthropologists. Although some (Eliot D. Chapple and others) have studied occupational groups, the field has largely been examined by sociologists. Anthropological literature on work in modern industrialized societies is limited.

Anthropologists have used their own methods in the study of occupational groups. Unlike many other approaches to occupational groups, anthropologists have used the concept of culture as a master concept to explain workers' behavior. This book has attempted to use the idea of culture as it relates to the technological base of the industry.

I have employed the anthropological method of participation in the researched population, using an emic perspective—that is, viewing the culture of construction workers from their frame of reference. This book is mainly an ethnographic account focusing on qualitative data gathered through observation and participation in construction projects.

Work and occupational role constitute focal points in the lives of people in an industrialized society. One's social and psychic well-being is to a large degree determined by one's occupational role. As anthropologists turn their attention increasingly to modern society, they will inevitably study occupational groups. As studies are developed by anthropologists on occupational groups and work environments, an anthropology of work may emerge as a subfield in the larger discipline of anthropology.

Epilogue/Women in construction: Something new

Every public construction project contains a section called "The Affirmative Action Program." Under this program, contractors are expected to hire and train minorities. Up until two years ago, affirmative action applied to blacks, Hispanics, American Indians, and Asians. After we received our contract on the Elm-Oak project, we were informed that affirmative action would also include women. Women were not classified as minorities, but they were made a part of the affirmative action programs in construction. The percentage of women we were supposed to hire was established at 3.5 percent of our working force.

When the men found out we had to hire women, it provoked a great deal of discussion. Viewpoints about the prospect of hiring women were far from uniform. There was much joking about the possibility of women working with the men 20 feet below in a trench, or working inside a pipe or high up on a scaffold. Many lunch time and barroom conversations on the subject took place. Here is a typical exchange that took place in the field office one afternoon, after work:

Donnie: Look, I ain't carrying no woman in my crew. I have a hard enough time carrying myself and making my costs. I can't be carrying no woman.

Chuckie: (with a twinkle in his eye and a smirky smile) I wouldn't mind having women working in the trenches with me.

Donnie: You degenerate. All you care about is grabbing a piece of ass. You got no brains. How you gonna make your costs if you gotta carry a woman?

Chuckie: Well, they could be flagmen. I mean flagladies. They could do that, couldn't they?

Donnie: If they want to work in construction they gotta do what everyone else does. I'd put 'em on a buster (a concrete breaker that is powered by compressed air). If they couldn't hack it, that's it, they're out.

Chuckie: Why you gotta be so pig-headed? If they can flag, why not use them? You gotta put a man on flagging anyway.

Mike: (a pipe foreman) They could do other jobs. I read where they had women on the Alaska pipeline, doing welding and surveying.

Chuckie: That's right, women could do surveying. They'd probably be good at it. They're more precise than men are. They use women in factories on precision work.

Kenny: (a bricklayer) Listen, my wife can lay brick as good as me.

Donnie: That ain't saying much.

Kenny: No shit. You should see the way she can lay up a brick wall. She can't lay block. They're too heavy for her. But laying brick, I'll put her up against anyone.

Donnie: Well, as far as I'm concerned, if a woman wants to work, she gotta do everything a man can. That includes running a buster, laying pipe, using a torch, using a shovel. Whatever we gotta do. I'm having trouble carrying myself. I can't carry no woman.

Chuckie: Ah, shut up! You said that already. You gotta adjust. It's new times now.

A year after we started the job, we hired our first woman on the project. She was hired to drive a large dump truck. We moved more than 100,000 cubic yards of dirt on this project, and we used at least ten trucks every day to move dirt or bring slag from the local steel mill for our road bed, or to deliver asphalt when we were paving streets. The woman's name was Mary. She was hired as a truck driver and was required to do what the other drivers did—move dirt and deliver slag or stone from the quarry. She was a petite woman, about five feet tall, with long black hair, a pretty face, and a shapely figure. She was feminine in every respect. Her youthful looks belied the fact that she had an 18-year-old daughter. In spite of her feminine appearance, some of the men falsely attributed lesbianism to her because she was after "a man's job." Mary was so small everyone wondered how she could reach the brakes and work the clutch on the huge truck she drove. But she had power steering, and she said she just moved the seat as far forward as it would go and that enabled her to reach all the pedals.

When she first went to work for us, she was definitely a novelty and attracted the men like flies. The superintendent grumbled that she was a disruptive influence. When she came into the field office in the morning with the rest of the men, there were lots of comments of "Ooops" and "I'm sorry," as the men caught themselves up short in their use of obscenities, with which they freely sprinkled all their conversations. It is strange that in spite of many other manifestations of male chauvinism, construction men never curse in front of women and are always gracious when in their presence. When they are by themselves they express all kinds of superior, sexist ideas and feelings.

For a while, after Mary came to the job, she was the subject of many salacious jokes. One day, one of the men remarked, "Hey, Mary's all dressed up today, she's wearing a bra." The men were constantly making jokes about Mary's truck, referring to the truck parts as if they were parts of Mary's body. The mechanics, in particular, thought they were being funny, in talking about "Mary's rear end," or the fact that "she needs a tune up," or "Mary needs her plugs and points changed," or that "her exhaust was plugged." The fact is that men who ran machines or drove trucks were referred to in the same way so that it was "Charlie's oil line blew" or "Joe has a flat tire" or "Wade's muffler is shot." There was the same identification of the man with the machine, referring to parts of the machine as if they were the man's. When this was done in Mary's case, it always evoked laughter.

Mary was well aware of the type of conversations and jokes that swirled around her, but she always maintained her aplomb and dignity. She could engage in repartee or ignore some silly remark, but she always insisted on being a full, participating member of the construction team on the job. When we organized a baseball game, Mary showed up at game time, sat in the miniature grandstand, and cheered for everyone. She drank beer, ate hot dogs and hamburgers, and swapped stories with the men about the job and construction work.

After a while, the men adjusted to Mary's presence and she was taken for granted as one of the construction drivers on the project. She was making $10.75 an hour, plus benefits, and she worked just as hard as the other drivers to earn her money. In the morning when she reported to work, Sal would bark out orders to her just as he did to the other drivers. During the day, her assignment might be switched several times, but she always responded quickly and efficiently. She showed up for work every day and never took any time off, even though once she broke her ankle and had to work in a cast. Another time, she backed into a parked car. No one criticized her, especially since some of the other men had similar incidents.

One of the high points for Mary on the job occurred when one of the Buffalo TV stations found out she was on the job and decided to do a story about her being the only female construction truck driver in the city. (Since that time, we have hired another woman, but Mary was the first.) TV reporters and camera men came to the job and followed Mary as she traveled over the job site and performed her duties. They interviewed several men on the job to ascertain their attitudes toward

Figure 15. Mary, our first woman truck driver, and her truck. (Photo courtesy of John Kaiser, Roy Crogan & Son.)

working with a woman. Tony Z., the grade foreman, expressed the general sentiment among the men when he said,

"When she first came here, she was a novelty and we all felt a little funny and uncomfortable when she was around. Now, she does her job and we accept her just like any other worker because she does her work and we don't treat her or think about her as someone special anymore."

The story about Mary was run on the evening TV news station, and all the men who happened to get into the TV picture, especially Tony, thanked Mary for making them "TV stars."

Appendix

TABLE 3. MODEL OF THE CONSTRUCTION INDUSTRY

Industry Characteristics	*Social Organization*	*Worker Behavior Patterns*
Uncertainty	Flexible	a. Acceptance of insecurity b. Crisis orientation c. Nonrepetitive work tasks
Hand tool technology	Decentralized decision making	a. Independence b. Control over the work process c. Autonomy d. Ownership of tools e. Loose supervision
Informal work system	Informal and variable rules and regulations Shifting roles and goals	a. Face-to-face relationships b. Friendship and kinship networks
Subcontracting	Diffusion and sharing of administration	a. Democracy among the trades b. Weak superior–subordinate relations
Localized nature of construction industry	Close-knit network of crafts, firms, and suppliers	a. Manifestation of traits of an occupational community b. Merging of work and nonwork activities

TABLE 4. CHARACTERISTICS OF AN OCCUPATIONAL COMMUNITY

Attributes	*Application to the Construction Industry*
1. Sense of group identity	Construction workers have identity by craft, union, and acquired skills.
2. Lifelong involvement	Long training period fosters lasting commitment to construction as an occupation. Job satisfaction encourages commitment.
3. Common values	Construction workers share: Independence and autonomy. Control over the work process. Deep involvement in work skills. High economic rewards and prestige. Ownership of tools and pride in craft.
4. Common definitions	Craft unions detail the role and rights of craftsmen and apprentices. Socialization defines roles of foreman and craftsmen. Building Trades Councils determine craft jurisdictions.
5. Common language within community	Each craft has its technical language. Each craft has its jargon for tools, methods, craft practices, materials, and dress, as well as jargon for construction work as a whole.
6. Community power over members	Union can bring members up on charges. Union can fine members and place them on probation. Expulsion from the union is tantamount to expulsion from trade.
7. Clear social limits of community	Community is composed of contractor's associations, worker's unions, architects and engineers professional societies, and networks of suppliers and equipment renters. Each local community has its own network.
8. Control over recruitment	Training and recruitment are controlled by craft union. Training schools are run by the union, not management. Recruitment is based on kinship and friendship.

Source: Adapted from William J. Goode, "Community within a Community," *American Sociological Review*, 22: 194–200.

Bibliography

Baumgartel, H., 1956, "Leadership Motivations and Attitudes in Research Laboratories." *Journal of Social Issues*, 12(2):24–31.

Becker, Howard S., and J. W. Carper, 1956, "Elements of Identification with an Occupation." *American Sociological Review*, 21:341–348.

Blauner, Robert, 1964, *Alienation and Freedom in American Industry*. Chicago: University of Chicago Press.

Blauner, Robert, 1966, "Work Satisfaction and Industrial Trends in Modern Society." In Reinhard Bendix and Seymour M. Lipset (eds.), *Class, Status, and Power*. New York: The Free Press.

Blum, Fred H. 1953, *Toward a Democratic Work Process*. New York: Harper.

Brayfield, A. H., and H. F. Rothe, 1951, "An Index of Job Satisfaction." *Journal of Applied Psychology*, 35:307–311.

Brooks, Thomas R., 1972, "Job Satisfaction: An Elusive Goal." *The American Federationist*, 79(10):1–7.

Bruyn, Severyn T., 1966, *The Human Perspective in Sociology*. Englewood Cliffs, N.J.: Prentice-Hall.

Cantril, Hadley, 1951, *Public Opinion, 1935–1946*. Princeton, N.J.: Princeton University Press.

Caplow, Theodore, 1954, *The Sociology of Work*. New York: McGraw-Hill.

Chao, Kang, 1968, *The Construction Industry in Communist China*. Hawthorne, N.Y.: Aldine.

Chapple, Eliot D. 1953, "Applied Anthropology in Industry." In A. L. Kroeber (ed.), *Anthropology Today*. Chicago: University of Chicago Press.

Cherry, Mike, 1974, *On High Steel*. New York: Quadrangle Press.

Chinoy, Ely, 1955, *Automobile Workers and the American Dream*. Garden City, N.Y.: Doubleday.

Civil Engineering, April 1979, pp. 63–68.

Colclough, J. R., 1965, *The Construction Industry of Great Britain*. London: Butterworths.

Colean, Miles L., and Robinson Newcomb, 1952, *Stabilizing Construction*. New York: McGraw-Hill.

Construction Equipment, "Employment Picture—1977." July 1977:22.

Construction Equipment, "Construction Employment—Bad News Today and Tomorrow." August 1977:11.

Cottrell, Fred W., 1951, "Death by Dieselization." *American Sociological Review*, 16.

Crozier, Michel, 1971, *The World of the Office Worker*. Chicago: University of Chicago Press.

Curle, Adam, 1949, "Incentive to Work: An Anthropological Appraisal." *Human Relations*, 2:41–47.

Drucker, Peter, 1950, *The New Society.* New York: Harper.

Durkheim, Emile, 1949, *The Division of Labor in Society.* New York: The Free Press.

Engineering News-Record, "Safety Net Result: Money Spent, Lives Saved," July 18, 1976:16.

Engineering News-Record, "Construction Volume," April 14, 1977:67.

Engineering News-Record, "Accident Record in Construction," November 15, 1979.

Fleishman, E. A., and E. F. Harris, 1962, "Patterns of Leadership Behavior Related to Employee Grievance and Turnover." *Personnel Psychology*, 15:43–46.

———, and H. E. Burtt, 1965, *Leadership and Supervision in Industry.* Columbus, Ohio: Ohio State University Bureau of Educational Research.

Foster, Charles, 1969, *Building with Men.* London: Tavistock.

Friedmann, E. A., and R. J. Havighurst, 1954, *The Meaning of Work and Retirement.* Chicago: University of Chicago Press.

French, J. R. P., Jr., J. Israel, and D. As, 1960, "An Experiment on Participation in a Norwegian Factory." *Human Relations*, 13:3–19.

Fromm, Erich, 1961, *Marx's Concept of Man.* New York: Frederick Ungar.

Gerstl, J. E., 1961, "Determinants of Occupational Community in High Status Occupations." *Sociology Quarterly*, 2:37–48.

Goldthorpe, John H., David Lockwood, Frank Bechhofer, and Jennifer Platt, 1968, *The Affluent Worker: Industrial Attitudes and Behavior.* London: Cambridge University Press.

Goode, William J., 1957, "Community Within a Community." *American Sociological Review*, 22:194–200.

———, 1967, "The Protection of the Inept." *American Sociological Review*, 32: 5–19.

Gouldner, Alvin W., 1954, *Patterns of Industrial Bureaucracy.* New York: The Free Press.

Gouldner, H. P., 1960, "The Norm of Reciprocity: A Preliminary Statement." *American Sociological Review*, 25:161–178.

Graves, Bennie, 1970, "Particularism, Exchange and Organizational Efficiency: A Case Study of a Construction Industry." *Social Forces*, 49:72–81.

———, 1974, "Conflict and Work Force Stability in Pipeline Construction." *Urban Life and Culture*, 2(4):415–431.

Haas, Jack, 1977, "A Study of High Steel Ironworkers' Reactions to Fear and Danger." *Sociology of Work and Occupations*, 4(2):147–171.

Haber, William, and Harold J. Levinson, 1956, *Labor Relations and Productivity in the Building Trades.* University of Michigan Bureau of Industrial Relations. Ann Arbor, Mich.: University of Michigan Press.

Hayes, Carleton J., 1946, *A Political and Cultural History of Modern Europe.* New York: Macmillan.

Heilbroner, Robert L., 1970, *The Economic Problem.* Englewood Cliffs, N.J.: Prentice-Hall.

Herzberg, F. et al., 1957, *Job Attitudes.* Pittsburgh, Pa.: Psychological Service of Pittsburgh.

Herzberg, Frederick, Bernard Mausner, and Barbara Snyderman, 1967, *The Motivation to Work.* New York: Wiley.

H.E.W. Report, 1975, *Work in America.* Cambridge, Mass.: The M.I.T. Press.

Highway and Heavy Construction, "Editorial." March 1978.

Hoppock, Robert, 1935, *Job Satisfaction.* New York: Harper.

H.U.D. Report, 1973, *Action Against Seasonal Unemployment in the Construction Industry.* Washington, D.C.: U.S. Government Printing Office.

Hughes, Everett C., 1959, *Men and Their Work.* New York: The Free Press.

———, 1959. The Study of Occupations. In Robert Merton, Leonard Broom, and Leonard Cottrell (eds.), *Sociology Today*. New York: Basic Books.

———, 1971, *The Sociological Eye: Selected Papers*. Hawthorne, N.Y.: Aldine.

International Labor Organization, 1969, *Construction Skills*. Geneva, Switz.: CIRF Publications.

Jacobson, E., 1951, *Foreman–Steward Participation Practices and Worker Attitudes in a Unionized Factory*. Ann Arbor, Mich.: University of Michigan Press.

Joyce, John T., 1973, "Construction Unions in the Seventies." *American Federationist*, 80:9–15.

Kasl, Stanislav V., 1974, "Work and Mental Health." In James O'Toole (ed.), *Work and the Quality of Life*. Cambridge, Mass.: The M.I.T. Press.

Katz, D., G. Gurin, and L. G. Floor, 1951, *Productivity, Supervision and Morale Among Railroad Workers*. Survey Research Center, Institute for Social Research. Ann Arbor, Mich.: University of Michigan Press.

———, N. Maccaby, and N. C. Morse, 1950, *Productivity, Supervision and Morale in an Office Situation*. Institute for Social Research. Ann Arbor, Mich.: University of Michigan Press.

Kay, E., J. R. P. French, and H. H. Meyer, 1962, "Behavioral Research Services Report, No. EBR-11." New York: The General Electric Co.

Kerr, W. A., 1958, "On the Validity and Reliability of the Job Satisfaction Index." *Journal of Applied Psychology*, 32:275–281.

Kluckhohn, Florence, 1940, "The Participant-Observer Technique in Small Communities." *American Journal of Sociology*, 46:331–343.

LeMasters, E. E., 1975, *Blue-Collar Aristocrats*. Madison, Wis.: The University of Wisconsin Press.

Levison, Andrew, 1974, *The Working Class Majority*. New York: Penguin Books.

Likert, R., 1961, *New Patterns of Management*. New York: McGraw-Hill.

Lipset, S. M., M. Trow, and J. Coleman, 1956, *Union Democracy*. New York: The Free Press.

Lopreato, Joseph, 1970, *Italian Americans*. New York: Random House.

Mattila, John M., and Erwin A. Gaumnitz, 1955, "An Econometric Analysis of Construction." *Bureau of Business Research and Service, Wisconsin Commerce Reports*. Madison, Wis.: University of Wisconsin Press.

Michigan Survey Research Center, 1971, *Survey of Working Conditions (November 1970)*. Ann Arbor, Mich.: University of Michigan Press.

Mills, C. Wright, 1953, *White Collar*. New York: Oxford University Press.

Mills, Daniel Q., 1972, *Industrial Relations and Manpower in Construction*. Cambridge, Mass.: The M.I.T. Press.

Mills, Ted, 1976, "Altering the Social Structure in Coal Mining." *Monthly Labor Review*, 99(10):3–10.

Monthly Labor Review, 94 (July 1971):7.

Moreno, J. L., 1937, *Who Shall Survive?* Washington, D.C.: Nervous and Mental Disease Publishing Co.

Morse, Nancy C., and E. Reimer, 1956, "The Experimental Change of a Major Organizational Variable." *Journal of Abnormal Social Psychology*, 52:120–129.

Myers, R. R., 1946, "Interpersonal Relations in the Building Industry." *Human Organization*, 5(Spring):1–7.

Norr, J. L., and L. L. Norr, 1978, "Work Organization in Modern Fishing." *Human Organization*, 37(2):163–171.

O'Brien, James J., and Robert G. Zilly, 1971, *Contractor's Management Handbook*. New York: McGraw-Hill.

Orton, E. S., 1976, "Change in the Skill Differential: Union Wages in Construction, 1907–1972." *Industrial and Labor Relations Review*, 30(October):6–24.

Otterbein, Keith F., 1977, *Comparative Cultural Analysis*, 2d ed. New York: Holt, Rinehart and Winston.

Park, Robert, 1968, "Human Nature, Attitudes and Mores." In C. Gordon and K. J. Gergen (eds.), *The Self in Social Interaction*. New York: Wiley.

Pilcher, William W., 1972, *The Portland Longshoremen*. New York: Holt, Rinehart and Winston.

Riemer, Jeffrey W., 1975, "On Building Buildings." Unpublished Ph.D. thesis, University of New Hampshire.

Roethlisberger, F. J., and W. J. Dickson, 1939, *Management and the Worker*. Cambridge, Mass.: Harvard University Press.

Ross, I. C., and A. Zander, 1957, "Need Satisfaction and Employee Turnover." *Personnel Psychology*, 10:327–338.

Scott, Rachel, 1974, *Muscle and Blood*. New York: Dutton.

Seidman, Joel, Jack London, Bernard Karsh, and Daisy L. Tagliacozzo, 1958, *The Worker Views His Union*. Chicago: University of Chicago Press.

Sheehan, George A. 1975, *Dr. Sheehan on Running*. Mountain View, Calif.: World Publications.

Slichter, Sumner H., J. J. Healy, and E. R. Livernash, 1960, *The Impact of Collective Bargaining on Management*. Washington, D.C.: Brookings Institute.

Smith, Patricia C., 1963, "Strategy for the Development of a General Theory of Job Satisfaction." *Cornell Studies of Job Satisfaction*. Ithaca, N.Y.: Cornell University Press.

Spier, John, 1959, *Elements of Job Satisfaction in the Railroad Operating Crafts*. Berkeley, Calif.: University of California Press.

Stinchcombe, Arthur, 1959, "Bureaucratic and Craft Administration of Production: A Comparative Study." *Administrative Science Quarterly*, 4:168–187.

Strauss, George, 1956, "Controls by the Membership in Building Trades Unions." *American Journal of Sociology*, 61(May):527–535.

———, 1958, "Unions in the Building Trades: A Case Study." *The University of Buffalo Studies*, 24(2):61–159.

———, 1963, "Professionalism and Occupational Associations." *Industrial Relations*, 2(2):7–31.

———, 1971, *Union Policies Toward the Admission of Apprentices*. Reprint No. 357. Berkeley, Calif.: University of California Press.

Taylor, Frederick W., 1911, *The Principles of Scientific Management*. New York: Harper.

Terkel, Studs, 1971, *Working*. New York: Pantheon Books.

Trist, E. L., and K. W. Bamforth, 1951, "Some Social Psychological Consequences of the Longwall Method of Coal-Getting." *Human Relations*, 4:3–38.

U.S. Bureau of the Census, 1967, *Census of Construction Industries*. Washington, D.C.: U.S. Government Printing Office.

———, 1976, *Census of Construction Industries*. Washington, D.C.: U.S. Government Printing Office.

U.S. Bureau of Labor Statistics, *Monthly Labor Review* (September 1967).

———, 1970, *Compensation in the Construction Industry*. Bulletin No. 1656. Washington, D.C.: U.S. Government Printing Office.

———, 1970, *Seasonality and Manpower in Construction*. Bulletin 1642. Washington, D.C.: U.S. Government Printing Office.

———, 1970, *Employment and Earnings*. Washington, D.C.: U.S. Government Printing Office.

U.S. Department of Commerce, 1970, *General Social and Economic Characteristics: Final Report*. Washington, D.C.: U.S. Government Printing Office.

———, 1976, *Construction Review*. Washington, D.C.: U.S. Government Printing Office.

U.S. Department of Labor, 1969, *Apprenticeship: Past and Present.* Washington, D.C.: U.S. Government Printing Office.

Van Zelst, Raymond H. 1952, "Empathy Test Scores of Union Leaders." *Journal of Applied Psychology*, 36:293–295.

———, 1952, "Validation of a Sociometric Regrouping Procedure." *Journal of Abnormal Social Psychology*, 47:299–301.

———, 1952, "Sociometrically Selected Work Teams." *Personal Psychology*, 5: 175–186.

Walker, C. R., and R. H. Guest, 1952, *Man on the Assembly Line.* Cambridge, Mass.: Harvard University Press.

Walker, Charles, 1962, *Modern Technology and Civilization.* New York: McGraw-Hill.

Wallick, Franklin, 1972, *The American Worker: An Endangered Species.* New York: Ballantine Books.

Wallman, Sandra, 1979, *Social Anthropology of Work.* London: Academic Press.

Weitz, J., and R. C. Nuckols, 1953, "The Validity of Direct and Indirect Questions in Measuring Job Satisfaction." *Personal Psychology*, 6:487–494.

Weschler, I. R., and R. E. Bernberg, 1950, "Indirect Methods of Attitude Measurement." *International Journal of Opinion Attitude Research*, 4:209–228.

Western New York, July 1977.

Whyte, William F., 1974, *Men at Work.* Westport, Conn.: Greenwood Press.

Wickert, F. R., 1951, "Turnover and Employees' Feelings of Ego-Involvement in the Day-to-Day Operations of a Company." *Personnel Psychology*, 4:185–197.

Woldman, Elizabeth, 1968, *Employment Status of School Age Youth.* Labor Force Report No. 11, U.S. Bureau of Labor Statistics, October. Washington, D.C.: U.S. Government Printing Office.

Recommended reading

Blauner, Robert, 1966, "Work Satisfaction and Industrial Trends in Modern Society." In Reinhard Bendix and Seymour M. Lipset (eds.), *Class, Status and Power.* New York: The Free Press.
This article is one of the best short summaries of the literature on work satisfaction and the various factors that are associated with the concept and the reality of job satisfaction in the United States.

Cherry, Mike, 1974, *On High Steel.* New York: Quadrangle Press.
A marvelous book, beautifully written and rich with the humor and humanity of the author, who was a construction worker and wrote about his work mates. While Mike Cherry deals with one trade, the ironworkers who erect structural steel, his book contains many insights about the construction industry in general. I have read this book many times and return to it often for the sheer pleasure of its truth and directness which reflect the world of construction workers.

Colean, Miles L., and Robinson Newcomb. 1952, *Stabilizing Construction.* New York: McGraw-Hill.
Although this book was written more than 25 years ago, it contains a sound and basic analysis of the characteristics and structure of the construction industry in the United States. It not only outlines the nature of the industry, but explains why, technologically, the construction industry is organized and administered the way it is.

Foster, Charles, 1969, *Building with Men.* London: Tavistock Publications.
This book contains a personalized account of what it is like to be in the construction industry on a day-to-day basis. The author was a small contractor in England, and he describes the human and technical problems of running a construction business. For me, the conversations in the book were very true to life.

Gerstl, J. E., 1961, "Determinants of Occupational Community in High Status Occupations." *Sociology Quarterly*, 2:37–48.
A good summary of the elements of an occupational community, of which construction workers are, in my view, one example.

Gouldner, Alvin W. 1954, *Patterns of Industrial Bureaucracy.* New York: The Free Press.
An excellent study of how bureaucratic methods of administration relate to

industrial workers. Gouldner compares and contrasts the differences in job satisfaction between workers who enjoy a large measure of autonomy and those subject to bureaucratic control.

Haber, William, and Harold J. Levinson, 1956, *Labor Relations and Productivity in the Building Trades.* University of Michigan Bureau of Industrial Relations. Ann Arbor, Mich.: University of Michigan Press.
This is another book which is quite old but which contains an excellent analysis of the characteristics of the construction industry and why it is organized the way it is.

Mills, Daniel Q., 1972, *Industrial Relations and Manpower in Construction.* Cambridge, Mass.: The M.I.T. Press.
This book is an up-to-date, overall summary of the nature of the construction industry.

Pilcher, William W., 1972, *The Portland Longshoremen.* New York: Holt, Rinehart and Winston.
This is an important book. It is one of the finest industrial ethnologies on workers in the United States. It is an excellent model for viewing certain occupational groups as occupational communities. The characteristics of the workers in this study are very similar to those found among construction workers.

Stinchcombe, Arthur, 1959, "Bureaucratic and Craft Administration of Production: A Comparative Study." *Administrative Science Quarterly*, 4:168–187.
A most insightful study of the contrast between the craft and the bureaucratic administration of work. Stinchcombe uses construction workers as his example of a craft occupation which enjoys autonomy and a nonbureaucratic mode of work administration. Stinchcombe sees construction workers as a professionalized work force enjoying many of the same benefits and satisfactions from work that professionals do.

Strauss, George, 1958, "Unions in the Building Trades: A Case Study." *The University of Buffalo Studies*, 24(2):61–159.
One of the best and most human studies of unions in the construction industry. It focuses on the business agent, who is the key union official closest to the construction workers in the field. The study was done by using the method of participant observation.

Wallman, Sandra, 1979, *Social Anthropology of Work.* London: Academic Press.
This is the only book of its kind, as far as I know. It is an entire book dealing with the anthropology of work. The book is the result of a colloqium on the subject, held in Bristol, England. It contains a series of studies on the anthropological concept of work as well as various studies of occupations in different parts of the world. The book seeks to develop a specific anthropological view of work and occupations.

Whyte, William F., 1974, *Men at Work.* Westport, Conn.: Greenwood Press.
Contains a series of studies of various occupations and how work is organized within those occupations. I found some of the studies that dealt with the crafts particularly illuminating, as they contain insights about craft occupations in general.

Case Studies in Contemporary American Culture, from Case Studies in Cultural Anthropology and Case Studies in Education and Culture, edited by George and Louise Spindler.

Applebaum, Herbert A. ROYAL BLUE: The Culture of Construction Workers
Construction workers' relationship to and attitudes toward their occupations.
1981/156 Pages/ISBN: 0-03-057309-2

Aschenbrenner, Joyce LIFELINES: Black Families in Chicago
Individual and family networks in an urban setting.
1975/160 Pages/ISBN: 0-03-012826-9

Collier, John, Jr. ALASKAN ESKIMO EDUCATION: A Film Analysis of Cultural Confrontation in the Schools (CSEC)
Classrooms in BIA schools; confrontation of Anglo and Eskimo cultures.
1973/130 Pages/ISBN: 0-03-088021-1

Daner, Francine Jeanne, THE AMERICAN CHILDREN OF KRSNA
A study of a contemporary alternative religion based on participant observation.
1976/128 Pages/ISBN: 0-03-013546-X

Davidson, R. Theodore CHICANO PRISONERS: The Key to San Quentin
Prison culture from perspective of the Family; the Baby Mafia.
1974/196 Pages/ISBN: 0-03-091616-X

Dougherty, Molly Crocker BECOMING A WOMAN IN RURAL BLACK CULTURE
A modified community study focusing on the social maturation of black adolescent girls in rural north Florida.
1978/128 Pages/ISBN: 0-03-014921-5

Friedland, William H./Nelkin, Dorothy MIGRANT: Agricultural Workers in America's Northeast
Work crew and its control, relationship with the outside world.
1971/281 Pages/ISBN: 0-03-085767-8

Gamst, Frederick C. THE HOGHEAD: An Industrial Ethnology of the Locomotive Engineer
Inside examination of railroad operations and their relationship to society at large.
1980/128 Pages/ISBN: 0-03-052636-1

Hicks, George L. APPALACHIAN VALLEY
Culturally distinctive folk culture of the Appalachians.
1976/128 Pages/ISBN: 0-03-077305-9

Hostetler, John A./Huntington, Gertrude E. CHILDREN IN AMISH SOCIETY: Socialization and Community Education (CSEC)
Amish school and community versus the outside world.
1971/119 Pages/ISBN: 0-03-077750-X

Hostetler, John A./Huntington, Gertrude E. THE HUTTERITES IN NORTH AMERICA
World view, technology, family and socialization, communal organization. New Fieldwork Edition
1980/119 Pages/ISBN: 0-03-065005-4

Jacobs, Jerry FUN CITY: An Ethnographic Study of A Retirement Community
An "active way of life" is designed but only a few participate.
1974/96 Pages/ISBN: 0-03-001936-2

Keiser, R. Lincoln THE VICE LORDS: Warriors of the Streets
Gang membership, territoriality, leadership in Chicago. New fieldwork edition.
1979/96 Pages/ISBN: 0-03-045396-8

Madsen William THE MEXICAN-AMERICANS OF SOUTH TEXAS, Second Edition
Anglo-American relations, class differences, folk beliefs, and acculturation. New edition with epilogue by Andre Guerrero.
1973/124 Pages/ISBN: 0-03-008431-8

McFee, Malcolm MODERN BLACKFEET: Montanans on a Reservation
Indian-oriented and white-oriented adaptation. Why assimilation has not occurred.
1972/134 Pages/ISBN: 0-03-085768-6

O'Toole, James WATTS AND WOODSTOCK: Identity and Culture in the United States and South Africa
Comparison of Watts, Los Angeles, and Woodstock, a Coloured ghetto in Cape Town.
1972/154 Pages/ISBN: 0-03-000936-7

Partridge, William L. THE HIPPIE GHETTO: The Natural History of a Subculture
Rituals, values, and sentiments; as a revitalization movement, not a counter culture.
1972/88 Pages/ISBN: 0-03-091081-1

Pilcher, William W. THE PORTLAND LONGSHOREMEN: A Dispersed Urban Community
Work culture, extra-work activities, union, family, and race relations.
1972/128 Pages/ISBN: 0-03-091289-X

Rosenfeld, Gerry "SHUT THOSE THICK LIPS!": A Study of Slum School Failure (CSEC)
Why and how inner city schools fail. The network of self-sustaining perceptions.
1971/120 Pages/ISBN: 0-03-085350-8

Spindler, George/Spindler, Louise URBAN ANTHROPOLOGY IN THE U.S.
A collection of case studies designed to give students exposure to four major segments of American urban society—*Chicano Prisoners: The Key to San Quentin; Lifelines: Black Families in Chicago; Fun City: An Ethnographic Study of a Retirement Community; Portland Longshoremen: A Dispersed Urban Community.*

Spindler, George/Spindler, Louise DREAMERS WITHOUT POWER: The Menomini Indians

Cognitive organization and adaptive strategies to the confrontation with Anglo-American culture and power in five contemporary groups.

1971/208 Pages/ISBN: 0-03-085542-X

Spindler, George/Spindler, Louise NATIVE NORTH AMERICANS: Four Cases

Four previously published CSCA covering most important culture areas of native North America—*Hano: A Tewa Indian Community in Arizona; The Kwakiutl: Indians of British Columbia; Modern Blackfeet: Montanans on a Reservation; Dreamers without Power: The Menomini Indians.*

1977/512 Pages/ISBN: 0-03-018401-0

Sugarman, Barry DAYTOP VILLAGE: A Therapeutic Community

Resocialization and values in a well-known drug rehabilitation center.

1974/134 Pages/ISBN: 0-03-086291-4

Ward, Martha C. THEM CHILDREN: A Study in Language Learning (CSEC)

How children in a small black Louisiana community acquire speech.

1971/99 Pages/ISBN: 0-03-086294-9

Williams, Melvin D. ON THE STREET WHERE I LIVED

An examination of the lifestyles of poor urban Blacks.

1981/160 Pages/ISBN: 0-03-056132-9

Wolcott, Harry F. THE MAN IN THE PRINCIPAL'S OFFICE: An Ethnography (CSEC)

Ethnography of middle class elementary school and principal. Shows how principal acts as mediator and system-supporter.

1973/334 Pages/ISBN: 0-03-091236-9